Quality Information, Quality Made
The Best in the Business™

PAL Publications

374 CIRCLE OF PROGRESS
POTTSTOWN, PA 19464
TEL: (610) 326-8597
FAX: (610) 326-4967
e-mail: palbooks@fcc.net
web site: www.palpublications.com

A Note to our Customers from

We've manufactured this book to the highest quality standards possible including the wrap around cover that is flexible, durable and water resistant to withstand the toughest on-the-job conditions.

PAL Publications has become the leader in providing the industrial trades with valuable, time saving information. All orders are processed in-house and shipped the next business day. We stand behind every book sold by providing exceptional service and customer satisfaction.

We Mean Business

AIR CONDITIONING

Testing/Adjusting/Balancing
A Field Practice Manual

Second Edition

John Gladstone

About the Author ...

Life Member and Fellow of ASHRAE, John has had over forty years experience in every phase of the industry including service, construction, planning, engineering and sales management. He holds a journeyman's card as well as a Master Mechanical Contractor's License and speaks from authoritative experience to journeyman mechanics, contractors and engineers alike. He was Chairman of the Examinations Committee of the Dade County Contractor's Examining Board (1965 – 1968) where he wrote some of the first contractor and journeyman examinations in the nation.

Mr. Gladstone graduated Miami-Dade Community College with honors, took his Bachelor's degree at St. Thomas and his Master's at Vermont College. He taught air conditioning design and license preparation at Miami-Dade Community College for many years.

He was a founding member of the Associated Air Balance Control (retired) and a Certificate Member of RSES (retired). Founder of the Air Conditioning Academy in Dania in 1977, Mr Gladstone continues to be active in educational and industry circles.

Information contained in this work has been obtained from sources believed to be reliable. However, neither Direct Brands, Inc. nor its authors guarantee the accuracy or completeness of any information published herein, and neither Direct Brands, Inc., nor its authors shall be responsible for errors, omissions, damages, liabilities or personal injuries arising out of the use of this information. This work is published with the understanding that Direct Brands, Inc. and its authors are supplying information but are not attempting to render engineering or other professional services. If such services are required, the assistance of an appropriate professional should be sought.

Preface

Here is the first handy air conditioning testing and balancing field manual that you can take to the job with you. Instant reference tables and charts critical to your work are at your fingertips; solutions to tough field problems are always within reach when you've got a copy of AIR CONDITIONING TESTING / ADJUSTING / BALANCING.

The latest techniques for testing, adjusting and balancing (TAB) air conditioning systems are presented in an easy-to-follow graphic format that includes charts, tables and illustrations specially designed by the author.

Our current need for energy conservation requires a greater emphasis on professional testing and balancing, because new heat recovery systems and conservation strategies rely on accurate feedback, testing and reporting. This book's coverage extends from standard balancing procedures to a complete trouble-shooting guide . . . from applying the fan laws and gas laws to estimating man-hour data for costing TAB activities and tasks. You'll also find out how to set duct pressure and water flow at the coil, correct for elevations and temperatures above standard, calculate and regulate air mixtures, and test exhaust hoods. To further save you time you'll discover how to adjust rpm and cfm without the use of instruments. In addition the author has included the basic formulas for heat flow, air flow, fans and pumps . . . a review of standard instruments and how to really use the Pitot tube . . . tips on gathering and recording field data . . . and much more. All of this essential information – previously scattered in many journals – is finally condensed into a single source to make your job easier.

The discussion is presented in a step-by-step format and stresses true understanding of environmental control through a knowledge of thermal and fluid flow as well as field pitfalls and problems.

AIR CONDITIONING TESTING / ADJUSTING / BALANCING meets the demands of a wide range of professionals in the cooling, heating and ventilating industries – including consultants, designers, project managers, estimators, contractors, plant engineers, utilities executives and inspectors. Mechanics will welcome the basic math review and teaching professionals will find the problem-and-solution style helpful in the classroom . . . all equations are presented in Systeme International D'Unites as well as U. S. Standard.

Introduction

The American Society of Heating, Refrigerating, and Air Conditioning Engineers (ASHRAE) defines system testing and balancing as "the process of checking and adjusting all the building environmental systems to produce design objectives." It uses these definitions:

Test: To determine quantitative performance of equipment.
Balance: To proportion flows within the distribution system (submains, branches, and terminals) in accordance with specified design quantities.
Adjust: To regulate the specified fluid flow rate and air patterns at the terminal equipment, (e.g., reduce fan speed, throttling, etc.)[1]

Definitions, of course, are necessary and while they are always enlightening they are seldom adequate when closely scrutinized and questioned. It is not our purpose here to debate these definitions but rather to arrive at a workable understanding of the testing, adjusting, and balancing (TAB) process in practical concretized situations. TAB is, after all, an affirmation of the design, construction, and installation of a total system's ability to meet its objectives, which is *the thermal comfort of the occupants of the treated space.*

The definition of thermal comfort according to ASHRAE is "that state of mind which expresses satisfaction with the thermal environment."[2] It is this *state of mind* that the system designer—and finally the TAB technician or engineer—must seek to satisfy. It follows, therefore, that many factors of both physiological and psychological parameters are compact to the tasks of the test and balance technician, and the professional TAB person must do considerably more than merely adjust and balance air quantities to meet the design.

Room air distribution quantities are calculated by the designer from the basic heat flow equation,

---

[1] *ASHRAE Handbook, Systems,* 1976, p. 40.1.
[2] *ASHRAE Comfort Standard,* 55–74, 2.11, p. 3. See also, *ASHRAE Handbook, Fundamentals,* 1977, p. 8.20.

$$\text{cfm} = \frac{\text{room sensible, Btuh}}{1.08 \ (\text{room db} - \text{supply air db})}$$

The supply air dry bulb temperature (db) is the air supplied to the room—or the air leaving the coil—sometimes called the leaving air db. That much is uncomplicated. But the room sensible heat gain—however simple and precise it may appear—is an ever-changing, mercurial quantity usually hovering in the penumbra of part load conditions.

The internal room load will vary with changes in occupancy, lighting levels, usage factors, etc. The load imposed on the structure from the outside will also vary with changes in outside air temperature, wind speed, and solar diffusion from cloud formation. As the space load changes, it must be balanced by the supply air designed to offset the load. This is accomplished by automatically varying the supply temperature at constant volume, varying the volume at constant temperature, or varying both temperature and volume. The TAB person has the responsibility of verifying this balance in the field.

Good design implies achieving thermal comfort—as defined above—with maximum energy efficiency and optimum operating conditions for extending the life of the equipment and minimum contamination of the interior/exterior environment. For a discussion of the variety of systems offered from which the designer may choose, the reader is referred to the *ASHRAE Handbook, Systems,* 1976, Section I, Chapters 1–20. The variable air volume system (VAV) appears to have many advantages. It would not be presumptuous to forecast an increased interest in VAV systems in the coming period on the part of researchers, manufacturers, and designers. Test and balance technicians and engineers will need to extend their experience in such systems and develop new techniques for field work. Meanwhile the observation should be emphasized that in those applications where hood exhaust systems are interfaced with variable air volume conditioning systems, air balancing may be extremely difficult. Designers and TAB people need to be particularly attentive to this problem.

The test, adjust and balance function is somewhat hindered by the fact that there is no universal standard balancing procedure. Unless the specification spells out the procedure, the technician may use any procedure that gives the best results for a particular system in a particular area. Whatever procedure is used, one point remains axiomatic; the air side should be balanced before the hydronic, steam, and refrigerant sides.

Contents

1

A Standard Balancing Procedure

In 1967, The Associated Air Balance Council published the first procedural standard for test and balance as part of National Standards for Field Measurements and Instrumentation. The following procedures are adapted from that document:

    I. All supply and return air duct dampers are set at full open position. All diffuser and side-wall grilles are set at full open position. Outside-air damper is set at minimum position.

        All controls are checked and set for full cooling cycle.

        Branch line splitter dampers are set to open position.

        All extractors and distribution grids are set in wide open positions.

    II. Drill all probe holes for static pressure readings, Pitot tube traverse readings, and temperature readings.

        Check motor electric current supply and rated running amperage of fan motors.

        Check fan and motor speeds.

        Check available adjustment tolerance.

   III. Make first complete air-distribution run throughout entire system, recording first-run statistics.

        Using Pitot tube traverse in all main ducts, branch ducts, and supply and return ducts, proportion all air in required amounts to the various main-duct runs and branch runs.

        Make second complete air-distribution run throughout entire system for check on proper proportion of air.

   IV. Using Pitot tube traverse, set all main-line dampers to deliver proper amount of cfm to all areas.

Using Pitot tube traverse, set all branch-line dampers to deliver proper amount of cfm to diffusers and side-wall supply grilles in each zone. Read cfm at each outlet and adjust to meet requirements.

Test and record all items as listed.

TESTING PROCEDURE, AIR SIDE

Test and adjust blower rpm to design requirements.

Test and record motor full-load amperes.

Make Pitot tube traverse of main supply ducts and obtain design cfm at fans.

Test and record system static pressures, suction, and discharge.

Test and adjust system for design recirculated air, cfm.

Test and adjust system for design cfm outside air.

Test and record entering air temperatures; dry-bulb heating and cooling.

Test and record entering air temperatures, wet-bulb cooling.

Test and record leaving air temperatures, dry-bulb heating and cooling.

Test and record leaving air temperatures, wet-bulb cooling.

Adjust all main supply and return air ducts to proper design cfm.

Adjust all zones to proper design cfm supply and return.

Test and adjust each diffuser, grille, and register to within 2% of design requirements.

Identify each grille, diffuser, and register as to location and area.

Identify each grille, diffuser, and register as to type and size, and include the manufacturers' ratings on all equipment.

Readings and tests of diffusers, grilles, and registers shall include required fpm velocity and test resultant velocity, as well as required cfm and test resultant cfm after adjustments.

All diffusers, grilles, and registers shall be adjusted to minimize drafts in all areas.

Future exhaust fans shall be tested and adjusted and set for requirements of future hoods. Installed fume hood fans shall be adjusted to cfm requirements as shown on plans and as specified. Tests of fume hood enclosures shall be made to determine velocities across opening in accordance with safety codes.

TESTING PROCEDURE, WATER SIDE

Open all valves to full position; close coil bypass stop valves; set mixing valve to full coil flow.

Remove and clean all strainers.

Examine water in system and determine if water has been treated and cleaned.

Check pump rotation.

Check expansion tanks to be sure they are not air-bound and the system is completely full of water.

Check all air vents at high points of water systems and be sure all are installed and operating freely.

Set all temperature controls so that all coils are calling for full cooling.

This should close all automatic bypass valves at coil and chiller. Follow same procedure when balancing hot-water coils; set on full call for heating.

Check operating of automatic bypass valve.

Check and set operating temperatures of boilers and chillers to design requirements.

Complete all air balancing before commencing waterside balancing.

BALANCING PROCEDURE, WATER SIDE

Set chilled-water and hot-water pumps to design gpm.

Adjust water flow through chiller.

Adjust water flow through boilers.

Check entering and leaving water temperatures through chillers and boilers, and set to correct design temperatures.

Balance all chilled-water and hot-water coils.

Upon completion of flow readings and adjustments at coils, tag and mark all settings and record data.

Recheck settings at pumps, chillers, and boilers and make necessary adjustments.

Install pressure gages on coil, read pressure drop through coil at set flow rate on call for full cooling and full heating. Set pressure drop across bypass valve to match coil full flow pressure drop to prevent unbalanced flow conditions when coils are on full bypass.

Repeat above procedure for bypass valve adjustment at chiller.

At each cooling and heating element, check and record the following:

1. Entering and leaving water temperatures.
2. Pressure drop across coil.
3. Pressure drop across bypass valve.

Check and record suction and discharge pressures and tdh at each pump.

List all pump nameplate data.

Record the running amperes and volts at each pump motor.

Record readings at all water-metering devices.

SOUND TESTS

Table 1 gives acceptable sound ranges for various applications. The sound level readings should be measured in decibels on the "A" and "C" scales of The

Table 1.

Type of Space	Recommended Noise Criterion Curve	Computed Equivalent Sound Level Meter Readings Weighing Scale-A dba
Broadcast studios	NC 15-30	25-30
Concrete halls	NC 20	30
Legitimate theaters (500 seats, no amplification)	NC 20-25	30-35
Music rooms	NC 25	35
Schoolrooms (no amplification)	NC 25	35
Conference rooms for 50	NC 25	35
Apartments and Hotels	NC 25-30	35-40
Assembly halls (amplification)	NC 25-30	35-40
Homes (sleeping areas)	NC 25-35	35-45
Conference rooms for 20	NC 30	40
Motion picture theaters	NC 30	40
Hospitals	NC 30	40
Churches	NC 30	40
Courtrooms	NC 30	40
Libraries	NC 30	40
Small private offices	NC 30-35	40-45
Restaurants	NCA 45	55
Colliseums for sports only (amplification)	NCA 50	60
Stenographic offices (typing and business machines)	NC 50	60
Factories	NCA 40-65	50-75

General Radio Company Sound Level Meter, or an equal instrument that meets the current ANSI Standard Z24.3-1944 based on the acoustic reference power of db/re 10.13 watts.

The readings should establish the total random sound level of the selected rooms or areas with the system in operation as compared to the total background sound level with the system not in operation. The system increase over the background level is recorded on the "A" and "C" scales in decibels. If the sound levels are above those listed in Table 1, then adjustments must be made to bring the level within the range set forth. Sound traps, insulation, or dampers may have to be added to the system by the contractor.

Sound levels should be measured approximately 5 ft above the floor on a line about 45 degrees from the center of the diffuser.

2

Basic Formulas for Air Flow

BASIC AIR FLOW EQUATION

The basic air flow equation through any opening, duct, grille, door, etc. is Q = AV—where Q is the air flow rate in cfm, A is the area of the opening in ft^2 and V is the air speed or velocity expressed in feet per minute. When presented graphically as the Air Circle Equation, it is easy to commit to memory.

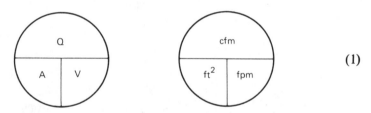

(1)

where any two variables are known, cover the unknown with your thumb to find the answer.

Example:

   Given: 2000 cfm passing through a duct 12 × 24 in.
   Find: The velocity, fpm
   Solution: If 12 × 24 in. = 288 in.^{2}, then $\dfrac{288 \text{ in.}^2}{144} = 2 \text{ ft}^2$

$$\frac{2000 \text{ cfm}}{2 \text{ft}^2} = 1000 \text{ fpm}$$

5

Example:

Given: A return air grille has a free net area of 1728 in.² and the velocity is measured with an anemometer at 500 fpm.

Find: The air volume, cfm

Table 2. Converting Round Duct Areas to Equivalent Square Feet Areas.

DUCT DIAMETER, INCHES	DUCT DIAMETER, mm	AREA FT²	AREA m²
8	203	0.3491	0.032
10	254	0.5454	0.051
12	305	0.7854	0.073
14	356	1.069	0.099
16	406	1.396	0.130
18	457	1.767	0.290
20	508	2.182	0.203
22	559	2.640	0.245
24	609	3.142	0.292
26	660	3.687	0.342
28	711	4.276	0.397
30	762	4.900	0.455
32	813	5.585	0.519
34	864	6.305	0.586
36	914	7.069	0.657
38	965	7.876	0.732
40	1016	8.727	0.811
42	1067	9.62	0.894
44	1119	10.56	0.981
46	1168	11.54	1.072
48	1219	12.57	1.168
50	1270	13.67	1.270
52	1321	14.75	1.370
54	1372	15.90	1.477
56	1422	17.10	1.586
58	1473	18.35	1.705
60	1524	19.63	1.824

Solution: $\dfrac{1728 \text{ in.}^2}{144} = 12 \text{ ft}^2$, then $12 \times 500 = 6000$ cfm

For round duct sizes Table 2 may be used for finding precalculated square areas quickly. For SI units, the Air Circle Equation is:

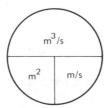

FINDING THE INFILTRATION OR VENTILATION AIR BY THE AIR CHANGE METHOD

$$\frac{\text{ft}^3 \times N_0}{60 \text{ min.}} = Q_0 \tag{2}$$

where
   Q_0 = outside air quantity, cfm
   N_0 = number of outdoor air changes per hour
   ft^3 = cubic contents of conditioned space

Example:

   Given: A room is $20 \times 20 \times 10$ ft.
   Find: The required cfm outside air for one air change per hour.
   Solution: $Q_0 = \dfrac{4000 \times 1}{60} = 66.6$ cfm

Example:

   Given: A room is $20 \times 20 \times 10$ ft.
   Find: The ventilation rate based on 2 air changes per hour.
   Solution: $Q_0 = \dfrac{4000 \times 2}{60} = 133$ cfm

FINDING THE NUMBER OF AIR CHANGES PER HOUR

$$N_t = \frac{Q_t}{\text{cfm/change}} \tag{3}$$

where

$$N_t = \text{total number of air changes per hour}$$
$$Q_t = \text{total air quantity, cfm}$$
$$\text{cfm/change} = \text{cubical contents/60 minutes}$$

Example:

Given: A room has a sensible heat of 120,000 Btuh. The temperature difference (rm db-lvg db) is 18 F. The room is 400 ft long, 15 ft wide and 10 ft high.

Find: The number of air changes per hour
Solution:

1. $\dfrac{120,000}{1.08 \times 18} = 6173$ supply air, Q_t

2. $\dfrac{60,000 \text{ ft.}^3}{60} = 1000$ cfm/air change/hour

3. $\dfrac{6173 \ Q_t}{1000 \text{ cfm/change}} = 6.173$ changes/hr, N_t

Example:

Given: A dry cleaning establishment is 400 ft long, 15 ft wide, and 10 ft high.
Find: The cfm based on an air change every 3 minutes
Solution: By rearranging Equation (3)

$$Q_t = \frac{\text{ft.}^3}{\text{min. per change}} = \frac{60,000 \text{ ft.}^3}{3} = 20,000 \text{ cfm}$$

MEASURING AIR FLOW THROUGH ORIFICE PLATES

The basic equation for air flow measurements through orifice plates is;

$$Q = 861 \ Kd^2 \ (h/w)^{1/2} \tag{4}$$

where
  Q = air flow, cfm
  K = discharge coefficient
  d = orifice diameter, ft
  h = pressure drop, in. WG
  w = weight of air, lb per ft^3
For standard air, w becomes 0.075

3

Basic Formulas for Fans

TO FIND THE THEORETICAL BRAKE HORSEPOWER (BHP)

$$bhp = \frac{cfm \times fan\ pressure}{6356 \times efficiency} \qquad (5)$$

Example:

A fan delivers 8000 cfm against 0.75 in. static pressure.
  Find: The brake horsepower if the efficiency is 63%.
  Solution: $\dfrac{8000 \times 0.75}{6356 \times 0.63} = \dfrac{6000}{4004} = 1.5\ hp$

Where the efficiency is not known a good rule-of-thumb is to allow 62.8% as the efficiency factor. Equation 5 becomes;

$$bhp = \frac{cfm \times P}{4000}$$

TO FIND THE FAN EFFICIENCY

Rearranging Equation 5,

$$Fan\ total\ efficiency = \frac{cfm \times P_t}{6356 \times bph}$$

$$Fan\ static\ efficiency = \frac{cfm \times P_s}{6356 \times bhp}$$

9

where
   P_t = total pressure
   P_s = static pressure

TO FIND THE TIP SPEED (TS)

$$TS = \frac{\pi D \times rpm}{12} = fpm \qquad (6)$$

where
   D = fan wheel diameter in inches
   TS = tip speed in feet per minute

Example:

A 24 in. diameter fan measures 850 rpm
   Find: The tip speed (TS)
   Solution: $\dfrac{3.14 \times 24 \times 850}{12} = 5338$ fpm

4

Belts, Pulleys, and Pulley Laws: Making the Adjustments

Drive sets for fans and blowers consist of a driver pulley on the motor shaft, a driven pulley on the blower shaft, and a belt or set of matched belts to transmit the power. Pulley formulas are usually given in pulley diameters; for accuracy they should be considered in actual pitch diameters. Figures 1, 2, and 3, and Table 3 give dimensions for standard variable sheaves.

The four basic pully laws are:

$$\text{rpm } P_f = \frac{\text{dia } P_m \times \text{rpm}}{\text{dia } P_f} \tag{7}$$

$$\text{rpm } P_m = \frac{\text{dia } P_f \times \text{rpm}}{\text{dia } P_m} \tag{8}$$

$$\text{dia } P_f = \frac{\text{dia } P_m \times \text{rpm}}{\text{rpm } P_f} \tag{9}$$

$$\text{dia } P_m = \frac{\text{dia } P_f \times \text{rpm}}{\text{rpm } P_m} \tag{10}$$

where
P_f = fan pulley or driven sheave
P_m = motor pulley or driver sheave

Figure 4 shows a Pulley Speed-O-Graph for rapid calculations of the pulley laws. Using this nomograph, the speed or size of either pulley can be determined when the other three factors are known.

1. Enter the chart from any given factor and follow the straight grid line to the point where it intersects, on the diagonal, the other given factor.

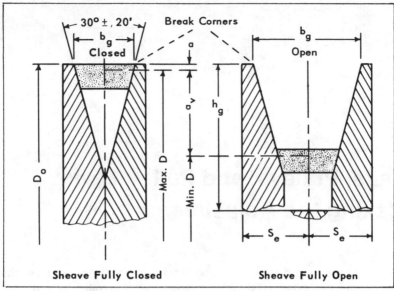

Figure 1.   Figure 2.

NOTE: D and d = Pitch diameter, large and small sheave
        D_O and d_O = Outside Diameter, large and small sheave

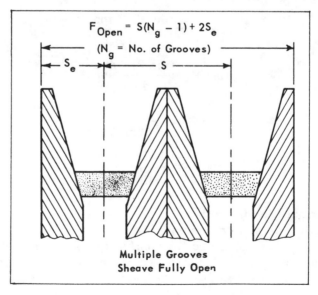

Figure 3.

Table 3. Variable Sheave Groove Dimensions.*

Cross Section	b_g Closed (Inches)	b_g Open (Inches)	h_g Minimum (Inches)	$2a$ (Inches)	$2a_v$ (Inches)	S_e Open Minimum (Inches)	S Minimum (Inches)
1430V	0.875 ± 0.005	1.582 ± 0.005	1.758	0.20	2.64	0.882	1.765
1930V	1.188 ± 0.005	2.142 ± 0.005	2.341	0.25	3.56	1.163	2.325
2530V	1.563 ± 0.007	2.823 ± 0.007	3.038	0.30	4.70	1.501	3.003
3230V	2.000 ± 0.007	3.665 ± 0.007	3.855	0.35	6.21	1.954	3.908
4430V	2.750 ± 0.007	5.132 ± 0.007	5.258	0.40	8.89	2.687	5.375

*Reprinted from *Engineering Standard*, Rubber Manufacturers Association, 1966, by permission of the authors.

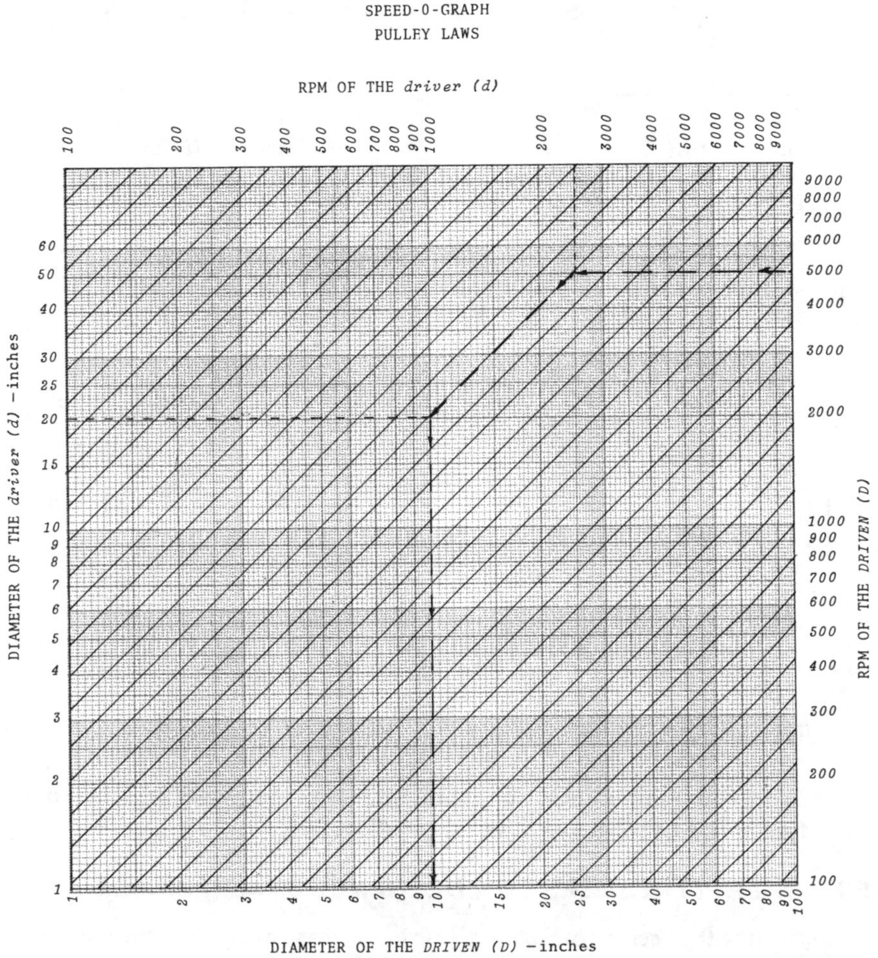

Figure 4. Speed-O-Graph: pulley laws.

Example 1. Given:
Diameter of Driver = 3 in.
DIAMETER OF DRIVEN = 12 in.
rpm of Driver = 5000
FIND: RPM OF DRIVEN

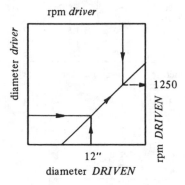

Example 2. Given:
Diameter of Driver = 30 in.
DIAMETER OF DRIVEN = 4 in.
RPM OF DRIVEN = 3750
FIND: rpm of Driver

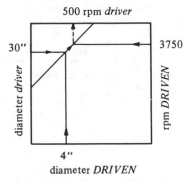

Example 3. Given:
rpm of Driver = 1000
RPM OF DRIVEN = 5000
Diameter of Driver = 10 in.
FIND: DIAMETER OF DRIVEN

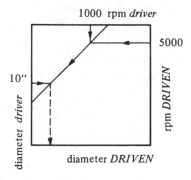

Example 4. Given:
rpm of Driver = 2500
RPM OF DRIVEN = 5000
Diameter of Driven = 10 in.
FIND: Diameter of Driver

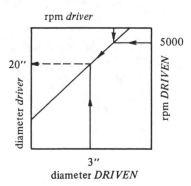

2. Follow the diagonal line to the point where it meets the third given factor.
3. From this point of intersection, move along the straight grid line to the fourth side of the margin for the solution.

FINDING RPM INCREASE OR DECREASE BY AMPERAGE

To determine the precent of rpm increase or decrease by reading the ammeter, the following formula applies:

$$amps_2 = amps_1 \left(\frac{rpm_2}{rpm_1}\right)^3 \tag{11}$$

Example:

A fan is turning 600 rpm and reading 20 amps. To deliver the proper cfm it is necessary to increase the fan speed to 700 rpm. Find the new amperage.

$$20 \times \left(\frac{700}{600}\right)^3 = 20 \times 1.588 = 31.75 \; amps_2$$

Table 4 gives calculated data for this equation. Using the above example, the increase in speed is 100 rpm; this is an increase of 100/600 or 16.66%. By

Table 4. Rpm Increase/Decrease.

(To determine the required change in fan speed multiply the measured amps by the given factor.)

% Rpm Increase	Multiply Amps By:	% Rpm Decrease	Multiply Amps By:
1		1	
2	1.06	2	0.94
3	1.09	3	0.92
4	1.13	4	0.88
5	1.16	5	0.86
6	1.19	6	0.83
7	1.23	7	0.80
8	1.26	8	0.78
9	1.30	9	0.75
10	1.33	10	0.73
11	1.37	11	0.70
12	1.40	12	0.68
13	1.44	13	0.66
14	1.48	14	0.64
15	1.52	15	0.61
16	1.56	16	0.59
17	1.60	17	0.57
18	1.64	18	0.55
19	1.69	19	0.53
20	1.73	20	0.51
21	1.77	21	0.49
22	1.82	22	0.47
23	1.86	23	0.45
24	1.90	24	0.44
25	1.95	25	0.42
30	2.20	30	0.34
35	2.46	35	0.28
40	2.75	40	0.22
45	3.05	45	0.17
50	3.38	50	0.12

interpolation the table shows that the original amps would have to be multiplied by 1.588, or 20 × 1588 = 31.75 amps.

FORMULAS FOR ADJUSTING SHEAVES

1. Given a change in cfm, find the new pulley setting.

$$pd_2 = \left(\frac{cfm_2}{cfm_1}\right) \times pd_1 \qquad (12)$$

Example:

Determine the new pitch diameter for 4000 cfm when a fan output is 3500 cfm at a 10 in. pitch diameter.

$$\left(\frac{4000}{3500}\right) \times 10 = 11.43 \text{ in.}$$

2. Given an increase in cfm, will the new brake horsepower overload the existing motor?

$$bhp_2 = \left(\frac{cfm_2}{cfm_1}\right)^3 \times bhp_1 \qquad (13)$$

Example:

Determine the new brake horsepower required to increase the cfm from 5000 to 5500 when the bhp is 0.8 and the motor is rated at 1 hp.

$$\left(\frac{5500}{5000}\right)^3 \times 0.8 = 1.06$$

Therefore, the motor needs to be changed to 1½ hp.

3. Given a maximum brake horsepower, find the new pitch diameter required to change from an existing pitch diameter.

$$pd_2 = \sqrt[3]{\frac{\text{max bhp}_2}{bhp_1}} \times pd_1 \qquad (14)$$

Example:

Determine the new pitch diameter to bring a 1 hp motor up to maximum when the present pd is 10 in. and bhp is 0.8.

$$\sqrt[3]{\frac{1}{0.8}} \times 10 \;=\; \sqrt[3]{1.25 \times 10} = 1.077 \times 10 = 10.77 \text{ pd}$$

where
  pd = pitch diameter
  bhp = brake horsepower
  cfm = air quantity at the fan

FORMULAS FOR FINDING BRAKE HORSEPOWER

The brake horsepower is the horsepower *actually* required to drive a fan; it includes the energy losses in the fan but does not include the drive loss between the motor and the fan. The bhp can only be determined by actual fan test in the field. By reading the actual running amperes and volts and plugging that into the formula, hp equals the square root of 3 times the product of volts, amps, power factor and motor efficiency divided by 746, the brake horse-power may be calculated. The difficulty with this formula, however, is finding the power factor and motor efficiency.

A more practical method of calculating the bhp is to construct a curve showing the amps vs. horsepower as illustrated in Figure 5.

Example:

A fan motor is rated at 2.4 amperes and $1\frac{1}{2}$ hp; a field test shows the running amps to be 1.9. Disconnecting the fan from the motor, the amps read 1.2. What is the brake horsepower?

  Step 1. Locate the no-load amps on the abscissa or amps scale (Point 1).
  Step 2. On the same scale, mark the $\frac{1}{2}$ no-load amps point (Point 2).
  Step 3. Mark the point where the nameplate amps and hp intersect (Point 3).
  Step 4. Draw a connecting line between points 2 and 3.
  Step 5. Plot the point on this line where it is intersected by $\frac{1}{2}$ the nameplate hp (Point 4).
  Step 6. Draw a smooth, solid curve through points 1 (1.2 amps), 3 (2.4 amps and 1.5 hp), and 4 (0.75 hp).
  Step 7. Locate the running amps on the abscissa (1.9) and move up the chart to the point of intersection on the solid curve to find the bhp. Brake horsepower = 1.2 (Point 5).

Constructing a curve in the field for this solution is not usually easy. Therefore, the fastest and simplest method for finding bhp is from the two formulas;

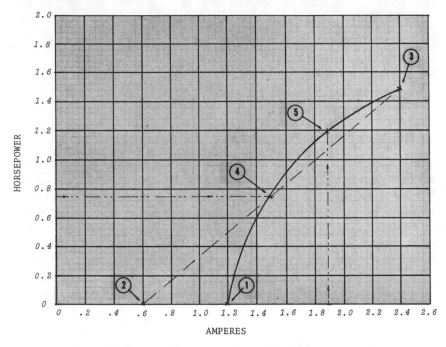

Figure 5. Amps vs Horsepower (to find brake horsepower.)

$$FA_a = \frac{V_n \times FA_n}{V_a} \tag{15}$$

where
    FA_a = actual full load amps
    FA_n = nameplate full load amps
    V_n = nameplate volts
    V_a = actual volts

Equation 15 establishes the actual or corrected full load amperes by reading the actual volts at the motor. Having determined FA_a, Equation 16 then solves the *approximate* bhp:

$$bhp = \frac{A_r - 0.5_{nl}}{FA_a - 0.5\,A_{nl}} \times hp_n \tag{16}$$

where
    A_r = field tested running amps
    A_{nl} = no load amps

FA_a = actual full load amps (corrected)

hp_n = nameplate horsepower

Using the same example as above and solving by formula rather than the *amps vs horsepower* curve:

$$bhp = \frac{\text{running amps} - \frac{1}{2} \text{ no load amps}}{\text{full load amps} - \frac{1}{2} \text{ no load amps}} \times \text{nameplate hp}$$

$$= \frac{1.9 - 0.6}{2.4 - 0.6} \times 1.5 = 1.08 \text{ approximate bhp}$$

V-BELTS

All standard V-belts are identified by a standard numbering system.

Single V-belts consist of a letter-numeral combination designating length of belt. An 8.0 in. belt is designated 2L080, a 52.0 in. belt is designated 3L520, and a 100.0 in. belt is designated 4L1000. The first digit indicates the number of digits in the inch length.

Variable speed V-belts consist of a standard numbering system that indicates the nominal belt top width in sixteenths of an inch by the first two numbers. The third and fourth numbers indicate the angle of the groove in which the belt is designed to operate. The digits following the letter "V" indicates the pitch length to the nearest 1/10 in. A belt numbered 1430V450 is a V-belt of 14/16 in. nominal top width designed to operate in a sheave of 30 degrees (groove angle) and have a pitch length of 45.0 in. Length tolerances for belts under 100 in. should be ±0.0025 over 100 in. ±0.0075.

The length of a V-belt is calculated by the formula:

$$L = 2C + 1.57(D + d) + \frac{(D - d)^2}{4C} \qquad (17)$$

where

C = Center distance between shafts, inches

D = Outside diameter of large sheave, inches

d = Outside diameter of small sheave, inches

The center distance between two shafts is given by the formula:

$$C = \frac{K^+\sqrt{K^2 - 32(D - d)^2}}{16} \qquad (18)$$

where

K = 4L − 6.28 (D + d)

L = Length of belt, inches

Table 5. Standard V-Belt Dimensions.*

STANDARD BELT LENGTHS

Standard Pitch Length Designation	Standard Effective Outside Length (Inches)				
	Cross Section				
	1430V	1930V	2530V	3230V	4430V
31.5	32.1				
33.5	34.1				
35.5	36.1	36.3			
37.5	38.1	38.3			
40	40.6	40.8			
42.5	43.1	43.3			
45	45.6	45.8			
47.5	48.1	48.3			
50	50.6	50.8	50.9		
53	53.6	53.8	53.9		
56	56.6	56.8	56.9	57.1	57.3
60	60.6	60.8	60.9	61.1	61.3
63	63.6	63.8	63.9	64.1	64.3
67	67.6	67.8	67.9	68.1	68.3
71	71.6	71.8	71.9	72.1	72.3
75	75.6	75.8	75.9	76.1	76.3
80		80.8	80.9	81.1	81.3
85		85.8	85.9	86.1	86.3
90		90.8	90.9	91.1	91.3
95		95.8	95.9	96.1	96.3
100		100.8	100.9	101.1	101.3
106		106.8	106.9	107.1	107.3
112		112.8	112.9	113.1	113.3
118		118.8	118.9	119.1	119.3
125			125.9	126.1	126.3
132				133.1	133.3

NOMINAL V-BELT CROSS SECTIONS

Cross Section	b$_b$ (Inches)	h$_b$ (Inches)
1430V	7/8	5/16
1930V	1-3/16	7/16
2530V	1-9/16	9/16
3230V	2	5/8
4430V	2-3/4	11/16

*Reprinted from *Engineering Standards*, Rubber Manufacturer's Association, 1966, by permission of the authors.

D = Outside diameter of large sheave, inches
d = Outside diameter of small sheave, inches
Table 5 gives the standard dimensions of variable pitch V-belts.

The belt speed for various rpm's may be found where the sheave pitch diameter is known, from the formula:

$$\text{fpm} = \text{pd} \times \text{rpm} \times 0.262 \tag{19}$$

Figure 6 gives a Belt Speed-O-Graph for rapid calculation of belt speeds in fpm where the pitch diameter and rpm is known. Enter at the abscissa for pitch diameter and move up the intersecting straight grid line in rpm; the diagonal intersect will read in fpm belt speed.

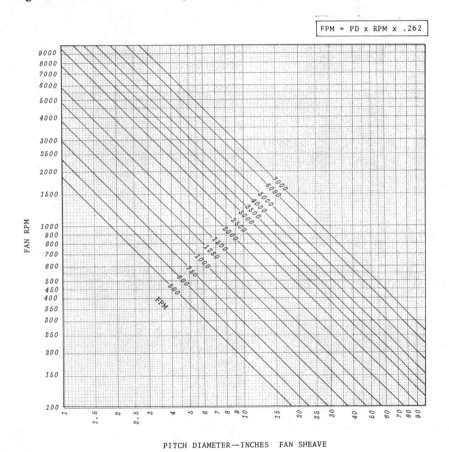

Figure 6. Speed-O-Graph: belt speeds.

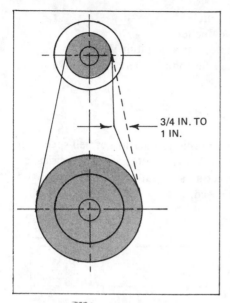

Figure 7. Belt Tension.

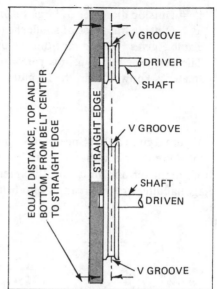

Figure 8. Pulley Alignment.

Belt tension adjustments should be made in three steps after the pulleys have been properly aligned.

Step 1. After initial installation of belts, run for five minutes and adjust for ¼ to 1 in. deflection as shown in Figure 7.

Step 2. Run for 12 hours and readjust tension.

Step 3. Run for 24 hours and make final adjustment for tension.

Only matched sets of pulleys should be used and accurate alignment of the driver and driven pulley should be made as shown in Figure 8.

5

Motors, Fans, and Fan Laws: Making the Adjustments

No universal procedure for testing (measuring) has yet been established, nor is there agreement as to the best method for balancing; but final adjustment at the fan leaves little room for methodological arguments. It is the fan laws, and the fan laws alone, that the field technician and engineer must rely upon to make adjustments. An understanding of the fan laws permits the field person to predict the effect of changes in fan speed or power or to determine in advance whether there is enough fan strength, i.e., fan size, pulley size, and horsepower.

Equations for the fan laws, pulley laws, and belt laws are found in many books and manuals, but they are often awkward to use or graphically unmanageable in field practice. The charts presented here have been designed to enable the technician and engineer to solve fan problems quickly and accurately without the use of a slide rule or a calculator. Another obvious benefit of the nomographic method is that it allows the operator to view a range of solutions simultaneously without reworking the equations. The charts shown are for the laws most commonly used in field work; they may, however, also be used for design purposes.

The fan laws are mathematical expressions that may be used to predict the performance of any fan when a fan of the same series has had established test data. For the test and balance technicians there are five basic fan laws that are absolutely essential to their work. With these formulas, the technician can adjust fan speeds to meet any new set of conditions.

The five basic fan laws are:

I. Cfm varies in direct proportion to rpm.

(a) $\dfrac{cfm_1}{cfm_2} = \dfrac{rpm_1}{rpm_2}$ 　　　　　　　　　　　　　(20)

(b) $cfm_2 = cfm_1 \times \dfrac{rpm_2}{rpm_1}$

(c) $rpm_2 = rpm_1 \times \dfrac{cfm_2}{cfm_1}$

II. Sp varies as the square of the rpm.

(a) $\dfrac{Sp_1}{Sp_2} = \left(\dfrac{rpm_1}{rpm_2}\right)^2$ 　　　　　　　　　　(21)

(b) $Sp_2 = Sp_1 \times \left(\dfrac{rpm_2}{rpm_1}\right)^2$

III. Hp varies as the cube of the rpm.

(a) $\dfrac{hp_1}{hp_2} = \left(\dfrac{rpm_1}{rpm_2}\right)^3$ 　　　　　　　　　　(22)

(b) $hp_2 = hp_1 \times \left(\dfrac{rpm_2}{rpm_1}\right)^3$

Example: A fan is operating at 0.5 in. static pressure at a speed of 1000 rpm and moving 2000 cfm of air at 1.5 horsepower. Using the above three formulas find the new Sp, rpm, and hp if the design calls for 2200 cfm.

1. $rpm_2 = \dfrac{cfm_2}{cfm_1} \times rpm_1 = \dfrac{2200}{2000} \times 1000 = 1100 \; rpm_2$

2. $Sp_2 = \left(\dfrac{cfm_2}{cfm_1}\right)^2 \times Sp_1 = \left(\dfrac{2200}{2000}\right)^2 \times 0.5 = 0.6 \; Sp_2$

3. $hp_2 = \left(\dfrac{cfm_2}{cfm_1}\right)^3 \times hp_1 = \left(\dfrac{2200}{2000}\right)^3 \times 1.5 = 2 \; hp_2$

IV. Amperage varies as the cube of the cfm.

(a) $\dfrac{amp_1}{amp_2} = \left(\dfrac{cfm_1}{cfm_2}\right)^3$ (23)

(b) $amp_2 = amp_1 \times \left(\dfrac{cfm_2}{cfm_1}\right)^3$

Example: A fan is handling 2000 cfm and drawing 2.5 amps, what will be the amperage reading necessary to bring the fan up to 2200 cfm?

$$amp_2 = amp_1 \times \left(\dfrac{cfm_2}{cfm_1}\right)^3 = \left(\dfrac{2200}{2000}\right)^3 \times 2.5 = 3.33\ amp_2$$

V. Horsepower varies as the square root of the pressure ratio cubed.

(a) $hp_2 = hp_1 \times \sqrt{\left(\dfrac{Sp_2}{Sp_1}\right)^3}$ (24)

Example: A fan is operating at 0.5 in. static pressure and 1.5 horsepower, what will the new horsepower be if the static pressure is increased to 0.6?

$$hp_2 = hp_1 \times \sqrt{\left(\dfrac{Sp_2}{Sp_1}\right)^3} = \sqrt{\left(\dfrac{0.6}{0.5}\right)^3} \times 1.5 = 2\ hp_2$$

Figures 9 thru 14 give Speed-O-Graphs for solving fan laws quickly and easily.

Fan and pulley laws can be arranged in a variety of ways and used for different purposes by design, application, and testing and balancing engineers. Here we discuss only a few of the basic laws as applied in the field.

Example:

Given: A system is operating at 0.5 in. static pressure, with a 1.5 hp motor and 800 rpm at the blower, and delivers 2000 cfm of air. The motor is rated at 1750 rpm and has a driven pulley of 4 in.

Find: The new rpm, static pressure, horsepower, and driver pulley to bring the system to 2800 cfm.

Solution: From Figure 4, the driven pulley has a pitch diameter of 8.75 in. or using Equation 7;

$$dia\ P_f = \frac{dia\ P_m \times rpm}{rpm\ P_f} = \frac{4 \times 1750}{800} = 8.75\ in.$$

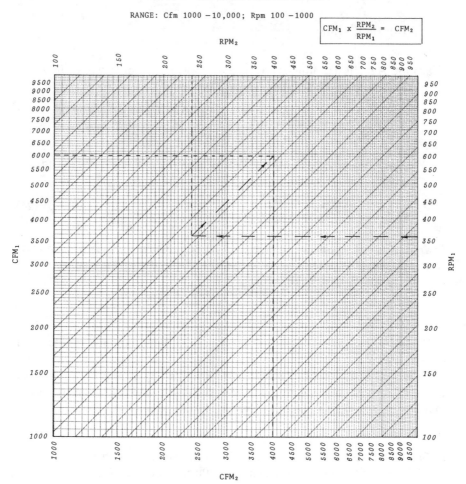

SPEED-O-GRAPH

FAN LAW NO.1   CFM VARIES DIRECTLY AS RPM

RANGE: Cfm 1000 −10,000; Rpm 100 −1000

$$CFM_1 \times \frac{RPM_2}{RPM_1} = CFM_2$$

FIGURE 9. Speed-O-Graph. *Given:* **360 rpm, 6000 cfm.** *Find:* **new cfm at 240 rpm. 1. Enter at 360 rpm₁ (right hand ordinate). 2. Move horizontally to the left to 240 rpm₂ (top of chart). 3. At intersection, move diagonally to the point at which 6000 cfm₁ intersects. 4. Drop down to the abscissa at bottom of chart to find the answer = 4000 cfm₂.**

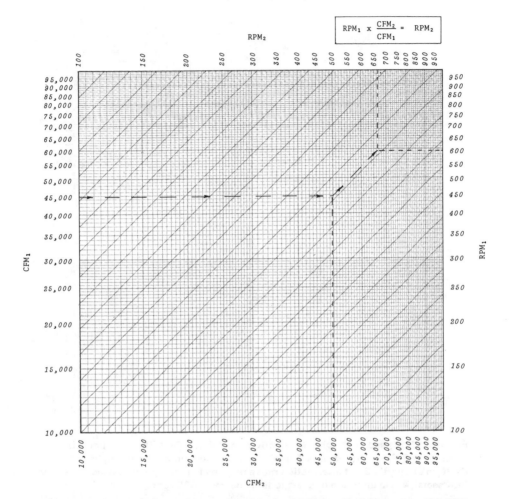

SPEED-0-GRAPH

FAN LAW NO.1   CFM VARIES DIRECTLY AS RPM

RANGE: Cfm 10,000 —100,000; Rpm 100 —1000

$$RPM_1 \times \frac{CFM_2}{CFM_1} = RPM_2$$

FIGURE 10. Speed-O-Graph. *Given:* **45,000 cfm, 600 rpm.** *Find:* **New rpm at 50,000 cfm. 1. Enter at 45,000 cfm₁ (left hand ordinate). 2. Move horizontally to the right to 50,000 cfm₂ (bottom of chart). 3. At the intersection, move diagonally to the point at which 600 rpm₁ intersects. 4. At top of chart find the answer = 665 rpm₂.**

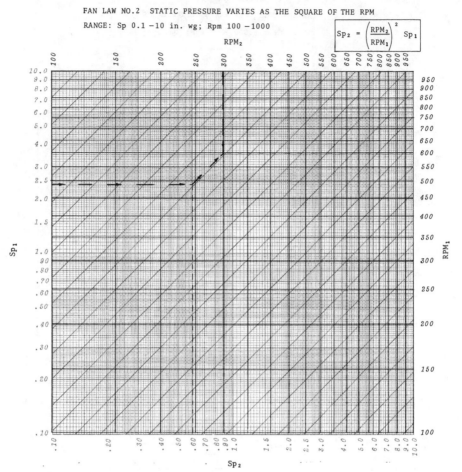

SPEED-0-GRAPH

FAN LAW NO.2   STATIC PRESSURE VARIES AS THE SQUARE OF THE RPM

RANGE: Sp 0.1 −10 in. wg; Rpm 100 −1000

$$Sp_2 = \left(\frac{RPM_2}{RPM_1}\right)^2 Sp_1$$

FIGURE 11. Špeed-O-Graph. *Given:* **600 rpm, 2.4 Sp.** *Find:* **The new rpm at 0.6 Sp. 1. Enter at 2.4 Sp (left hand ordinate). 2. Move horizontally to the right to 0.6 Sp (bottom of chart). 3. At intersection move diagonally to the point at which 600 rpm intersects. 4. Go up to top of chart to find answer = 300 rpm₂.**

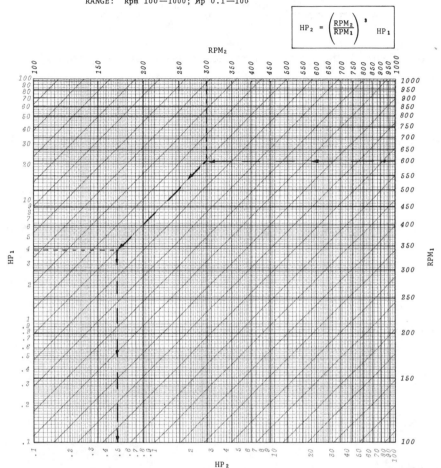

SPEED-0-GRAPH

FAN LAW NO.3 HORSEPOWER VARIES AS THE CUBE OF THE RPM
RANGE: Rpm 100—1000; Hp 0.1—100

$$HP_2 = \left(\frac{RPM_2}{RPM_1}\right)^3 HP_1$$

FIGURE 12. Speed-O-Graph. *Given:* **600 rmp at 4 hp.** *Find:* **The new hp at 300 rpm. 1. Enter at 600 rpm₁ (right ordinate). 2. Move horizontally to the left to 300 rpm₂. 3. At the intersection, move diagonally to the point at which 4 hp intersects. 4. Drop down to the abscissa at bottom of chart to find the answer = 0.5 hp₂.**

From Figure 10, the new rpm is 1120, or using Equation 20;

$$rpm_2 = rpm_1 \times \frac{cfm_2}{cfm_1} = 800 \times \frac{2800}{2000} = 1120 \text{ } rpm_2$$

From Figure 11, the new static pressure is 0.098, or using Equation 21;

$$Sp_2 = Sp_1 \times \left(\frac{rpm_2}{rpm_1}\right)^2 = 0.05 \times \left(\frac{1120}{800}\right)^2 = 0.098$$

From Figure 12, the new horsepower is 4.116, or using Equation 22;

$$hp_2 = hp_1 \times \left(\frac{rpm_2}{rpm_1}\right)^3 = 1.5 \times \left(\frac{1120}{800}\right)^3 = 4.116$$

From Figure 4, the new pitch diameter of the driver pulley can be found, 5.6, or using Equation 10;

$$\text{dia } P_m = \frac{\text{dia } P_f \times rpm}{rpm \text{ } P_m} = \frac{8.75 \times 1120}{1750} = 5.6 \text{ in.}$$

Obviously, the increase in the motor pulley size will require a new motor pulley and a change of belt:

$$\frac{5.6\pi}{2} - \frac{4\pi}{2} = 8.79 - 6.28 = 2.5 \text{ in.}$$

The new belt must be 2.5 inches longer.

Example:

   Given: A system is circulating 4500 cfm and is drawing 4 amps actual current. The motor is rated at 1725 rpm and 7.3 amps and has a pulley pitch diameter of 5 in. The pitch diameter of the fan pulley measures 12 in. Field measurements indicate that the system is critically short of air.

   Find: The maximum amount of air that the existing fan can accommodate without changing the motor and new pitch diameter of the driver pulley needed to make the adjustment.

   Solution: From Figure 4, the fan speed is found to be 719 rpm, or using Equation 7;

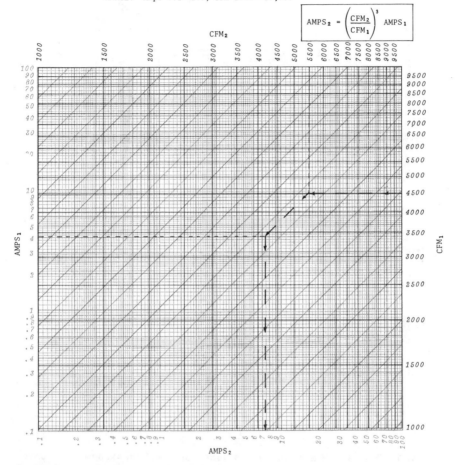

SPEED-O-GRAPH

FAN LAW NO.4   AMPERAGE VARIES AS THE CUBE OF THE CFM

RANGE: Amps 0.1—100; Cfm 1000—10,000

$$AMPS_2 = \left(\frac{CFM_2}{CFM_1}\right)^3 AMPS_1$$

FIGURE 13. Speed-O-Graph. *Given:* **4500 cfm at 4 amps.** *Find:* **The new amps at 5500 cfm. 1. Enter at 4500 cfm₁ (right ordinate). 2. Move horizontally to the left to 5500 cfm₂. 3. At the intersection, move diagonally to the point at which 4 amps intersects. 4. Drop down to the abscissa at bottom of chart to find the answer = 7.3 amps₂.**

FÁN LAW NO.5   HORSEPOWER VARIES AS THE SQUARE ROOT OF THE PRESSURE RATIO CUBED

RANGE: Hp 0.1—100; Sp 0.1—10

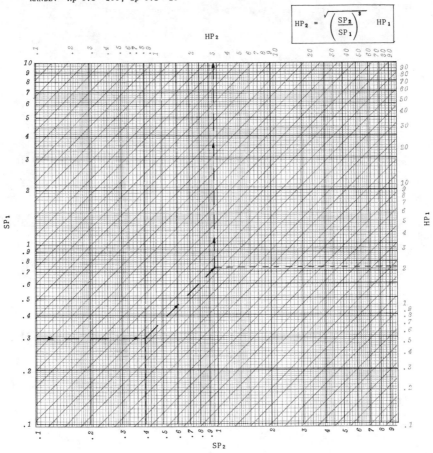

FIGURE 14. Speed-O-Graph. *Given:* **0.3 static pressure at 2 horsepower.** *Find:* **The new horsepower at 0.4 static.   1. Enter at 0.3 Sp_1 (left hand ordinate).   2. Move horizontally to the right to 0.4 Sp_2 (bottom of chart).   3. At intersection move diagonally to the point at which 2 hp_1 intersects.   4. Go up to top of chart to find answer = 3 hp_2.**

$$\text{rpm } P_f = \frac{\text{dia } P_m \times \text{rpm}}{\text{dia } P_f} = \frac{5 \times 1725}{12} = 719 \text{ rpm}$$

The new cfm can be found using Figure 13, and arriving at 5500 cfm$_2$, or by Equation 23;

$$\text{amps}_2 = \text{amps}_1 \times \left(\frac{\text{cfm}_2}{\text{cfm}_1}\right)^3$$

or

$$\text{cfm}_2 = \text{cfm}_1 \times \sqrt[3]{\frac{\text{amps}_2}{\text{amps}_1}} = 4500 \times 1.222 = 5500 \text{ cfm}_2$$

Using the new data, 4500 cfm, 5500 cfm$_2$, and 719 rpm, the new fan speed can be found by using Figure 10, or by Equation 20 (c):

$$\text{rpm}_2 = \text{rpm}_1 \times \frac{\text{cfm}_2}{\text{cfm}_1} = \frac{5500}{4500} \times 719 = 878 \text{ rpm}_2$$

Returning to Figure 4, it can be seen from the given 1725 rpm motor speed, 878 rpm fan speed and 12 in. fan pulley pitch diameter that the required new motor pulley pitch diameter will be 6.1 in., or by Equation 10.

$$\text{dia } P_m = \frac{12 \times 878}{1725} = 6.1 \text{ in.}$$

Now, assuming that the system is operating with a 2 hp motor at 0.3 in. water gage (WG) static pressure, and the design dictates that the system operate at 0.4 in. WG, will the motor have to be changed to bring the static pressure up another 0.1 in.? Because Fan Law No. 5 states that horsepower varies as the square root of the pressure ratio cubed, Figure 14 will show that the motor will have to be increased by one horsepower to accommodate the increase of 0.1 in. WG in static pressure, or by Equation 24;

$$\text{hp}_2 = \text{hp}_1 \times \sqrt{\left(\frac{\text{Sp}_2}{\text{Sp}_1}\right)^3} = 2\sqrt{\left(\frac{0.4}{0.3}\right)^3} = 3.079 \text{ hp}_2$$

Because the horsepower requirements fall on the hairline, it may require that a slight compromise in static pressure be made somewhere in the system to avoid the problem of going beyond a 3 hp motor, but the motor service factor may allow for this small difference.

These charts can be used to solve a variety of system problems and to reduce the number of man-hours required for testing and balancing. Reworking the above examples several times will provide the necessary facility to use the charts with speed and ease. A little practice sliding a pair of 60 degree angles across the page will simplify the procedure.

6

Using the Fan Laws to Balance and Adjust Without the Aid of Instruments

Field testing, balancing, and adjusting air systems requires a complete set of quality instruments. The capital investment is substantial and some of these instruments demand periodic calibration and maintenance. Yet, it is possible, under some conditions, to adjust such systems without the use of instruments at all. To do so requires a firm understanding of the basic pulley laws and fan laws.

Adjustable pulleys usually have a range of five turns between the fully open and fully closed positions. Therefore, one complete turn of the sheave will change the pitch diameter 0.2 in. and the speed of the driven pulley (P_f) by about 50 rpm. For a typical 1725 rpm driven pulley (P_m) with a 4 in. pitch diameter (pd) and an 8 in. pd driven pulley (P_f), the blower speed would be

$$\text{rpm } P_f = \frac{\text{dia } P_m \times \text{rpm}}{\text{dia } P_f} \tag{7}$$

substituting,

$$\frac{4 \times 1725}{8} = 862 \text{ rpm}$$

If the pitch diameter of the motor pulley is changed by taking 4.5 turns up, the new pitch diameter would become

$$4.5 \times 0.2 = 0.9 \text{ in.; and 4 in. } + 0.9 = 4.9 \text{ } P_m. \text{ Therefore,}$$

35

$$P_f \text{ rpm} = \frac{4.9 \times 1725}{8} = 1057 \text{ rpm}_2$$

Checking:

1. The average rpm per turn is 50. Therefore,

$$50 \times 4.5 = 225 \text{ rpm}$$

2. The average inches per turn is 0.2. Therefore,

$$0.2 \times 4.5 = 0.9 \text{ in.}$$

3. And 862 rpm + 225 rpm = 1087 rpm

The final check, therefore is less than 3% off, which is acceptable.

Figure 9 gives a chart for the first fan law (Equation 20), which states that the cfm varies in direct proportion to the rpm. Combining the fan laws with the pulley laws under actual field conditions, it is easily possible to adjust to final air quantity using no tools but a set of good thermometers.

Example: A system is observed to be operating as follows: driver pulley (P_m) pitch diameter is 4 in. fully open; motor name plate is 1750 rpm; driven pulley (P_f) pitch diameter is 8 in.; temperature drop across the coil (Δt) is 24 F. The system must be adjusted to correct design air quantity for a 19 F Δt as specified.

Solution: Assuming that the heat transfer difference between cfm_1 and cfm_2 is negligible, two equations can be constructed:

$$cfm_1 = \frac{Btuh}{(1.08)(\Delta t)_1}$$

$$cfm_2 = \frac{Btuh}{(1.08)(\Delta t)_2}$$

or

$$\frac{cfm_1}{cfm_2} = \frac{(\Delta t)_2}{(\Delta t)_1} \tag{25}$$

since cfm varies in direct proportion to the rpm as stated in Equation 20, then,

$$\frac{rpm_1}{rpm_2} = \frac{(\Delta t)_2}{(\Delta t)_1} \tag{26}$$

or

$$rpm_2 = rpm_1 \frac{(\Delta t)_2}{(\Delta t)_1}$$

First, determine the rpm of the driven pulley from the chart in Figure 4, 875 rpm. This datum could also be measured by tachometer or found by Equation 7:

$$\frac{4 \text{ in.} \times 1750 \text{ rpm}}{8 \text{ in.}} = 875 \text{ rpm}$$

Then, substituting for Equation 26:

$$rpm_2 = 875 \text{ rpm} \times \frac{24 \text{ F}}{19 \text{ F}} = 1105 \text{ rpm}_2$$

Once the new rpm requirement is known, the new pitch diameter can be found using Figure 4 or by Equation 10:

$$pd_2 = \frac{8 \times 1105}{1750} = 5.05 \text{ in.}$$

The increase in air quantity to bring the temperature drop to 19 F therefore requires a new driver pulley pitch diameter of 5 in., or an increase of 1.00 in. to step up the speed of the driven pulley an additional 230 rpm. Since the pitch diameter was 4 in. the fully open pulley can be closed about 5 turns. See above for check figures.

Of course, the pulley size may be too small to accomodate the necessary increase, or the motor may not be able to handle the additional load, in which case the motor or pulley would have to be changed. If the pulley is increased from 4 in. pd to, say, 5 in. pd, an increase in belt length would also be required. The new belt size may be calculated from:

$$\frac{\pi(pd)_1}{2} - \frac{\pi(pd)_2}{2} \tag{27}$$

or one half the difference in the circumference between the first and second pulleys. Substituting for Equation 27,

$$\frac{5\pi}{2} - \frac{4\pi}{2} = 7.85 - 6.28 = 1.57 \text{ in.}$$

The new belt must be 1.57 inches longer.

7

Basic Instruments and Their Use

The basic portable instruments required for field use are:

1. *Speed counter* for measuring rpm.
   a. Tachometer
   b. Stroboscope
   c. Revolution counter
2. *Clamp-on volt–amp meter* for measuring voltage and amperes.
3. *Thermometers and psychrometers* for measuring duct and space temperatures and relative humidity. Testing thermometers for ducts and coils should be 15 in. laboratory type glass, calibrated in 1/10 degree calibrations.
4. *Rotating vane anemometer* for measuring air flow at a coil face or large return air grille. This instrument is calibrated in feet and must be used with a timing instrument, such as a stop watch, to measure air flow in feet per minute. Taking care not to block the air flow, the technician must pass the instrument across the entire face of the coil so as to get an average reading in feet of air over a timed period. The fpm (V) multiplied by the square surface (A) = the cfm (Q). If the free area of the coil or grille can not be determined from the manufacturer's catalog, it must be accurately measured.
   The *rotating vane anemometer* is a propeller type instrument and requires periodic calibration as well as careful handling to maintain accuracy. In addition, it must be used with a correction factor curve.
5. *Swinging vane anemometer* for measuring air flow at a grille or diffuser. This is a more rugged instrument than the rotating vane and has the further advantage of reading in direct fpm. However, because the air

leaving a grille or diffuser is never of a uniform velocity, the technician must take several readings to find the average velocity. Ceiling diffusers are usually read at four points and averaged out. Large side wall grilles and return air grilles present a greater problem and here, the rotating vane serves better.

Using the *swinging vane anemometer* on large grilles, the technician must establish traverse points across the face of the grille, 6 inch divisions are best, and then take a number of readings at each point to determine the average velocity.

6. *Pitot tube* to probe the air inside of a duct. The *Pitot tube* is the real workhorse for the test and balance technician. It is used in conjunction with an inclined manometer to read the velocity pressure (Vp), static pressure (Sp), and total pressure (Tp) of a fluid or gas inside of a duct. A detailed discussion of the use of this instrument follows below.

7. *Inclined manometer* to measure air pressure inside of a duct. This instrument—also known as a draft gage, air meter, air velocity meter, or air gage—is a simple, foolproof device, which responds directly to the air pressure exerted against it (transmitted from the Pitot tube), and reads directly in inches of water. Ranges for these instruments vary, and the technician should have one or more instruments to cover the range of 0 to 5 in. water. This will read for air from 400 to 9000 fpm.

For low velocity work, a *micromanometer* or hook-gage is needed. This instrument is marked in scales of thousands of an inch and is used widely for readings at hoods, perforated ceilings, high precision industrial work, and test laboratories. The extreme accuracy of this instrument provides a standard against which other instruments can be calibrated.

8. *Magnahelic pressure gage* to measure air pressure inside of a duct. "Magnahelic" is not a generic term but is registered by Dwyer Instrument Company; it is used here as a convenience. The magnahelic gage is a diaphragm operated gage that has several advantages over a liquid manometer: 1. It need not be leveled to zero and can be used easily on a ladder or unlevel surface. 2. When hooked up to the Pitot tube it need not be purged of air bubbles as the liquid manometer may. 3. There is less chance of parallax error in reading the dial face. 4. It is easily transported without the chance of losing the liquid charge. Unless extreme accuracy is required, this instrument may replace the manometer for average air conditioning work, and like the manometer, is available in a variety of ranges. The dial is only 4 in. diameter and therefore has a limited scale; several instruments are required to cover the normal ranges encountered in average air conditioning jobs.

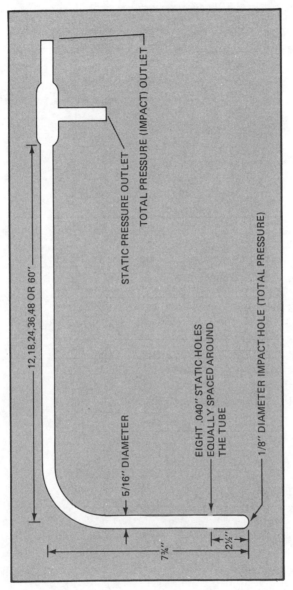

Figure 15. Standard Pitot Tube.

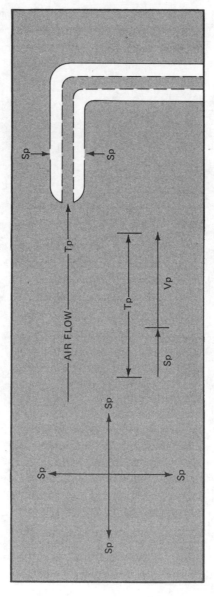

Figure 16. How Pressure is exerted on a Pitot Tube.

MEASURING DUCT PRESSURES AND FLOW RATES

During the last couple of years some manufacturers have developed and placed on the market a factory constructed "balancing station" which, when installed along with the duct system as a permanent appurtenance may eliminate the need for Pitot tube traverses and indicate the internal conditions at a glance. But the application of such "balancing stations" is by no means widespread thus far, and the Pitot tube traverse remains the only reliable, practical method of air balancing.

A Pitot tube is essentially a precisely calibrated stainless steel tube within a tube. The inner tube has a 1/8 in. diameter hole at the top; this is the impact hole which when positioned against the air flow in a duct, will sense the total pressure. The outer tube has eight small holes drilled around it—these are static pressure holes; the outer tube is the static tube. Figure 15 shows a standard 5/16 in. dia. Pitot tube; Figure 16 is an exaggerated view of a Pitot tube, showing how it functions inside of the duct. The inner tube is sometimes called the *impact tube*, and sometimes called the *total pressure tube*.

MAKING THE DUCT TRAVERSE

If the velocity of the air stream were uniform, then one reading at any point would be sufficient. But the air moving along the duct wall loses speed because of friction, consequently—assuming no special turbulence motion—the velocity in the duct center will always be greater. Since the velocity pressure is seldom uniform, a series of velocity pressure readings must be taken over an equal pattern across the duct section. Figure 17 shows recommended Pitot tube locations for traversing round and rectangular ducts.

For round ducts, *the tangential method* is the most common traverse. The duct is divided into n zones of equal area by concentric circles of radii, R_1, R_2, R_3, etc. and 20 readings are taken along two diameters. On paper, this may seem confusing but in practice it is quite uncomplicated. Two holes need to be drilled in the duct at cross diameters and the Pitot tube carefully marked— with a China marking pencil, or small strips of duct tape—for ten positions in each hole, giving 20 readings.

Table 6 gives precalculated measurements for a ten point traverse for round duct, by marking the Pitot tube accordingly, it may be inserted into the duct hole and a pressure reading taken at each mark. Following the same procedure across the other diameter will give a total of 20 readings. Figure 18 shows how to mark a Pitot tube for a 20 in. round duct using Table 6 to obtain the measured markings.

For square or rectangular ducts, a minimum of 16 readings must be taken at

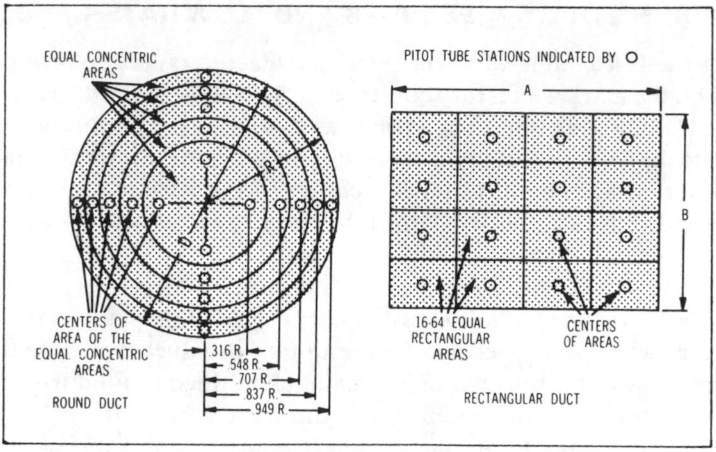

Figure 17. Pitot Tube Traverse Cross-section.

Table 6. Calculated Ten Point Pitot Tube Traverses For Round Ducts From 12 To 40 in.*

Traverse Point Number		1	2	3	4	5	6	7	8	9	10
Multiplier; Distance from Inside Wall to Pitot Point		.025	.083	.146	.225	.342	.647	.774	.855	.918	.975
Pipe Diameter	INCHES FROM INSIDE WALL TO TRAVERSE POINT										
12		⅜	1	1¾	2¾	4⅛	7⅞	9¼	10¼	11	11⅝
13		⅜	1⅛	2	3	4½	8½	10	11	11⅞	12⅝
14		⅜	1⅛	2	3⅛	4¾	9¼	10⅞	12	12⅞	13⅝
15		⅜	1¼	2¼	3⅜	5⅛	9⅞	11⅝	12¾	13¾	14⅝
16		⅜	1¼	2⅜	3⅝	5½	10½	12⅜	13⅝	14¾	15⅝
18		½	1½	2⅝	4⅛	6⅛	11⅞	13⅞	15⅜	16½	17½
20		½	1⅝	2⅞	4½	6⅞	13⅛	15½	17⅛	18⅜	19½
22		⅝	1¾	3¼	5	7½	14½	17	18¾	20¼	21⅜
24		⅝	2	3½	5½	8¼	15¾	18½	20½	22	23⅜
26		⅝	2⅛	3¾	5⅞	8⅞	17⅛	20⅛	22¼	23⅞	25⅜
28		¾	2¼	4⅛	6⅜	9⅝	18⅜	21⅝	23⅞	25¾	27¼
30		¾	2½	4⅜	6¾	10¼	19¾	23¼	25⅝	27½	29¼
32		¾	2⅝	4⅝	7¼	11	20¾	24¾	27⅜	29⅜	31¼
34		⅞	2⅞	5	7⅝	11⅝	22	26¼	29	31¼	33⅛
36		⅞	3	5¼	8⅛	12¼	23⅝	27⅞	30¾	33	35
38		1	3⅛	5½	8½	13	24½	29⅜	32½	34⅞	37
40		1	3⅜	5⅞	9	13¾	25⅞	31	34¼	36¾	39

*These calculated inches for marking the Pitot Tube for a ten station traverse are worked to the nearest eighths. For ducts other than listed, use the multiplier. All figures shown are in inches distance to the inside wall.

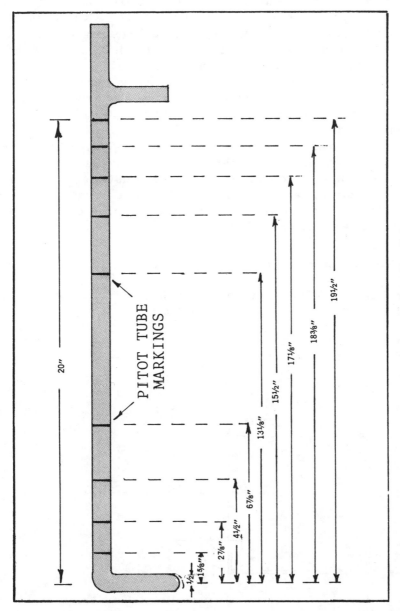

Figure 18. Marking the Pitot Tube.

centers of equal square or rectangular areas. Most text books and manuals agree that no area should exceed 36 sq. in.; that is 6 × 6, 7 × 5 1/2, 7 1/2 × 4 7/8, 8 × 4 1/2 in., etc. But actual field experience leads the author to believe that for average air conditioning work careful readings in areas up to 45 sq. in. is permissible. Naturally, if the readings seem at all odd or suspicious then the traverse must be based on smaller areas.

Four holes must be drilled in one wall of the duct in such a manner as to establish equal areas and the Pitot tube must be marked with the same measurements so that four different positions will be read at each hole giving a total of 16 readings.

Figure 19 illustrates a hole pattern for a 16 × 16 in. square duct with the Pitot tube marked and inserted in the hole for a reading at the 12th station.

For a four point traverse, the first hole is drilled at a distance of 1/8 the duct width away from the wall. The second hole is drilled at a distance of 1/4 the duct width away from the first hole. The third and fourth holes are the same distance apart as the second and first hole.

Example: If duct is 16 × 16 in. drill:

Hole No. 1 1/8 × 16 = 2 in. from duct wall
Hole No. 2 1/4 × 16 = 4 in. from hole No. 1

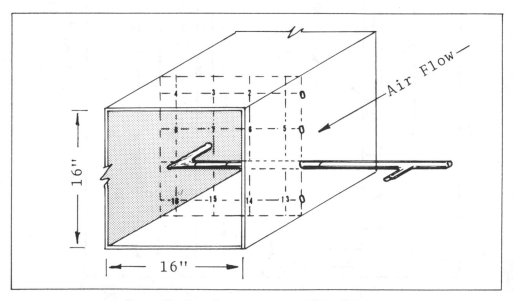

Figure 19. Duct Hole Pattern for 16 Point Traverse.

Table 7. Calculated Four Point Pitot Tube Traverses For Rectangular Ducts From 12 Through 29 In.

DUCT LENGTH OR WIDTH INCHES	INCHES FROM WALL TO 1ST POINT	INCHES FROM 1ST POINT TO 2ND POINT	INCHES FROM 2ND POINT TO 3RD POINT	INCHES FROM 3RD POINT TO 4TH POINT	INCHES FROM 4TH POINT TO FAR WALL
12	1.50	3.00	3.00	3.00	1.50
13	1.625	3.25	3.25	3.25	1.625
14	1.750	3.50	3.50	3.50	1.750
15	1.875	3.75	3.75	3.75	1.875
16	2.00	4.00	4.00	4.00	2.00
17	2.125	4.25	4.25	4.25	2.125
18	2.250	4.50	4.50	4.50	2.250
19	2.375	4.75	4.75	4.75	2.375
20	2.50	5.00	5.00	5.00	2.50
21	2.625	5.25	5.25	5.25	2.625
22	2.750	5.50	5.50	5.50	2.750
23	2.875	5.75	5.75	5.75	2.875
24	3.00	6.00	6.00	6.00	3.00
25	3.125	6.25	6.25	6.25	3.125
26	3.250	6.50	6.50	6.50	3.250
27	3.375	6.75	6.75	6.75	3.375
28	3.50	7.00	7.00	7.00	3.50
29	3.625	7.25	7.25	7.25	3.625

Hole No. 3 1/4 × 16 = 4 in. from hole No. 2
Hole No. 4 1/4 × 16 = 4 in. from hole No. 3
Hole No. 4 is   2 in. from far wall

Total = 16 in.

If the Pitot tube is marked with the same measurements, there will be 16 readings at the centers of equal squares.

For a five point traverse, the first hole is drilled at a distance of 1/10 the duct width and the succeeding holes are 1/5 the duct width apart. A six point traverse requires the first hole to be drilled 1/12 the distance of the duct width and the succeeding holes are 1/5 the duct width apart. Tables 7, 8, and 9, give precalculated measurements for 4, 5, and 6 point traverses for square and rectangular ducts and the same method is used to mark off the Pitot tube.

In Figure 19, the measurements are shown as distance in inches away from the Pitot tip—or duct wall—to conform with Table 6 for round ducts. But when marking off for square or rectangular ducts as shown in Tables 7–9, the measurements are made in distance between holes; the Pitot tube must be marked the same way.

Some Points to Remember When Making the Pitot Tube Traverse

1. Locate the Pitot tube in a clear, straight duct section providing at least 8 diameters upstream and 2 diameters downstream of the Pitot tube free of elbows, transitions or reductions.
2. Make accurate measurements and take careful readings. The Pitot tube traverse takes time, do not try to rush it; there are no short-cuts.
3. The standard Pitot tube is 5/16 in. diameter, drill probe holes 9/16 in. diameter to accomodate without chafing. When finished, cover holes with snap buttons and square of duct tape. If duct has internal liner, plug hole with a #5 bottle cork.
4. Check the impact and static holes regularly for plugging. Particularly with insulated ducts, these holes will occasionally plug up.
5. Check tubing—especially at connection ends—for leaks.
6. Check gage for zero reading at level before each set of traverses. Keep the manometer level.

PITOT TUBE—DRAFT GAGE HOOK-UPS

Figure 20 illustrates a standard manometer with a range of 0–2 in. water. When air pressure is imposed on the high pressure tube connection, the liquid level on the scale is pushed down, forcing the liquid column to rise on the low pressure side. The low pressure side of the gage may be referred to as the

Table 8.   Calculated Five Point Duct Traverses For Rectangular Ducts From 27 Through 36 In.

DUCT LENGTH OR WIDTH INCHES	INCHES FROM WALL TO 1ST POINT	INCHES FROM 1ST POINT TO 2ND POINT	INCHES FROM 2ND POINT TO 3RD POINT	INCHES FROM 3RD POINT TO 4TH POINT	INCHES FROM 4TH POINT TO 5TH POINT	INCHES FROM 5TH POINT TO FAR WALL
27	2.7	5.4	5.4	5.4	5.4	2.7
28	2.8	5.6	5.6	5.6	5.6	2.8
29	2.9	5.8	5.8	5.8	5.8	2.9
30	3.0	6.0	6.0	6.0	6.0	3.0
31	3.1	6.2	6.2	6.2	6.2	3.1
32	3.2	6.4	6.4	6.4	6.4	3.2
33	3.3	6.6	6.6	6.6	6.6	3.3
34	3.4	6.8	6.8	6.8	6.8	3.4
35	3.5	7.0	7.0	7.0	7.0	3.5
36	3.6	7.2	7.2	7.2	7.2	3.6

Table 9.   Calculated Six Point Duct Traverses For Rectangular Ducts From 32 Through 44 In.

DUCT LENGTH OR WIDTH INCHES	INCHES FROM WALL TO 1ST POINT	INCHES FROM 1ST POINT TO 2ND POINT	INCHES FROM 2ND POINT TO 3RD POINT	INCHES FROM 3RD POINT TO 4TH POINT	INCHES FROM 4TH POINT TO 5TH POINT	INCHES FROM 5TH POINT TO 6TH POINT
32	2.66	5.33	5.33	5.33	5.33	5.33
33	2.75	5.50	5.50	5.50	5.50	5.50
34	2.83	5.66	5.66	5.66	5.66	5.66
35	2.91	5.83	5.83	5.83	5.83	5.83
36	3.00	6.00	6.00	6.00	6.00	6.00
37	3.08	6.16	6.16	6.16	6.16	6.16
38	3.16	6.33	6.33	6.33	6.33	6.33
39	3.25	6.50	6.50	6.50	6.50	6.50
40	3.33	6.66	6.66	6.66	6.66	6.66
41	3.41	6.83	6.83	6.83	6.83	6.83
42	3.50	7.00	7.00	7.00	7.00	7.00
43	3.58	7.16	7.16	7.16	7.16	7.16
44	3.66	7.33	7.33	7.33	7.33	7.33

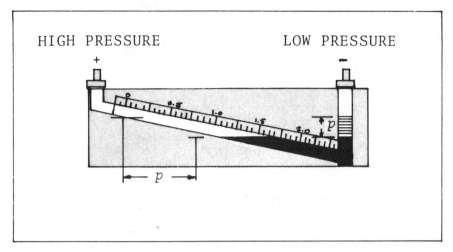

Figure 20. Draft Gage.

negative side, and the high pressure side as the positive side. This distinction is important in understanding the *Pitot tube–draft gage hook-up*.

Pitot tube hook-ups for various conditions are shown in Figures 21–23. It may be seen from these figures that whatever the condition, the hook-up for reading velocity pressure remains unchanged. The impact tube goes to the high pressure side and the static tube goes to the low pressure side.

When taking a static pressure reading in a supply duct, the impact tube is disconnected and the static tube is connected to the *high side*. To read static pressure in an exhaust duct, the static tube is connected to the *low side*.

To read total pressure in a supply duct, the static tube is disconnected and the impact tube is connected to the *high pressure* side of the gage. This connection is the same for reading total pressure in an exhaust duct when the total pressure is positive. But when the total pressure is negative the connection must be made to the *low side* of the gage. As mentioned earlier, total pressure is seldom considered in air conditioning and ventilation work.

Figure 24 shows the hook-up used to measure resistance across a filter or coil. Manufacturers of air filters, cooling and condenser coils, and similar equipment often publish data from which approximate air flow can be determined. It is characteristic of such equipment to cause a pressure drop which varies proportionately to the square of the flow rate. Along with the hook-up in Figure 24 an air flow vs resistance curve is shown. This is a straight line curve because it is plotted on a logarithmic scale. Using the manufacturer's data, a reading of 0.50 in. of water would indicate a flow of 2000 cfm. Pressure drop measurements of this kind are taken with two static tips; the impact tube is not used.

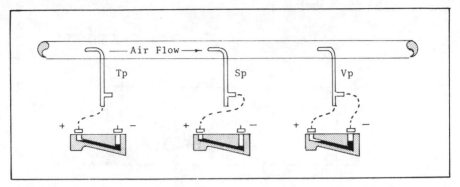

Figure 21. Supply Duct—Total Pressure Is Positive.

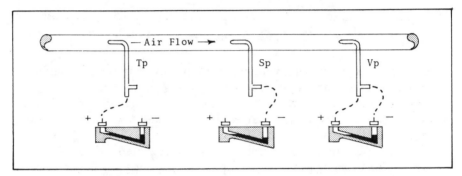

Figure 22. Exhaust Duct—Total Pressure Is Positive.

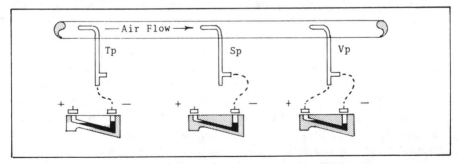

Figure 23. Exhaust Duct—Total Pressure Is Negative.

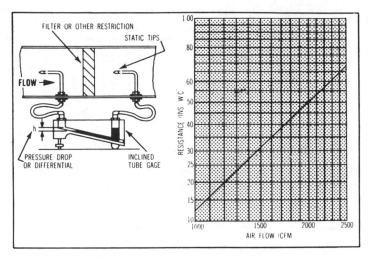

Figure 24. Filter Drop Hook-up.

Figure 58 offers a Speed-O-Graph for rapid solution of total pressure in relation to static, velocity, and velocity pressure.

Fan pressures may also be measured with a manometer. The fan *total pressure* is the difference between the total pressure at the fan outlet and the total pressure at the fan inlet, therefore two impact tubes are used in the hook-up as shown in Figure 25. The pressure difference on the manometer measures the total mechanical energy that the fan adds to the air.

Fan *static pressure* is the fan total pressure less the fan velocity pressure and is measured with the static tube on the discharge of the fan and the impact tube on the intake of the fan. By subtracting the total pressure at the intake from the static pressure at the discharge the fan static may be calculated. Figure 26 illustrates the hook-up.

USING THE ROTATING VANE ANEMOMETER

Figure 27 illustrates a coil being measured to calculate the square feet of actual opening before running a traverse with a rotating anemometer. In Figure 28 the balance technician is performing the anemometer travese to get a reading in feet. To find feet per minute (fpm) flow this instrument must be maneuvered over a timed course; that is to say, the rotating anemometer must be used with a stop watch or similar timing instrument as discussed earlier.

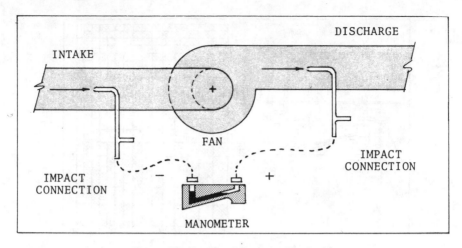

Figure 25. Fan Total Pressure Hook-up.

The formula for air flow is:

$$cfm = \frac{ft_m\,A\,F}{2} \qquad (28)$$

where
   ft_m = measured anemometer reading, feet
   A = free face area of grille, ft^2

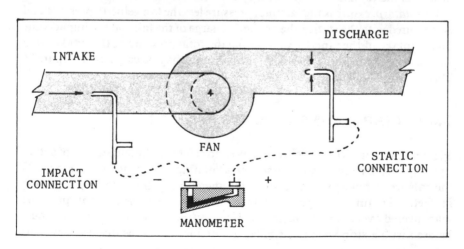

Figure 26. Fan Static Pressure Hook-up.

Figure 27. Measuring a coil to calculate the square feet of opening prior to running a traverse with the anemometer.

F = instrument correction factor

2 = two minute timed averaging pass across face of coil.

It has been found that, a two minute timed traverse gives better averaging accuracy across the coil face or return air grille than a one minute pass. It is recommended that two or more two minute traverses be made across the air stream and then averaged. Because of the friction drag in the instrument wheel, the rotating vane anemometer requires periodic calibration as well as the use of a correction factor. Each instrument is furnished with an F curve. Figure 29 gives values of F for this instrument.

The balancing technician in Figure 28 should not be standing in front of the coil face but rather to one side of the coil. Most anemometers come

Figure 28. Traversing a coil with anemometer and stop watch to measure the average velocity flow across the face.

equipped with extension handles for reaching out to perform the traverse without obstructing the air flow. Some controversy notwithstanding the rotating vane anemometer remains one of the most practical instruments for measuring total air flow at coils, return air grilles, and side wall grilles—if used correctly.

USING THE DEFLECTING VANE ANEMOMETER

In Figure 30 the test and balance technician is shown using an Alnor Velometer to average out velocity. Although this instrument does not have to be used with a stop watch or timer, it is not as accurate for this particular purpose as is the rotating vane anemometer. It is a cumbersome instrument for traversing large areas and the technican has no alternative but to stand directly in front of the air stream thereby altering the air flow pattern and upsetting the accuracy of the reading.

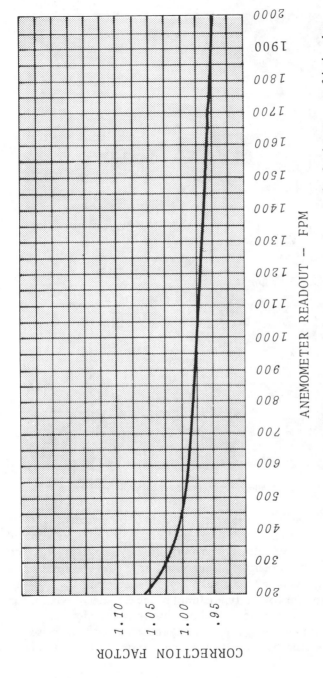

Note: This curve, adapted from the U.S. Bureau of Standards, corrects the velocity for instrument friction loss. Manufacturer's outlet K factors or grille free area factors are separate factors in addition to the above.

Figure 29. Anemometer Correction Curve For Standard 4 in. Rotating 8 Blade Biram Anemometer.

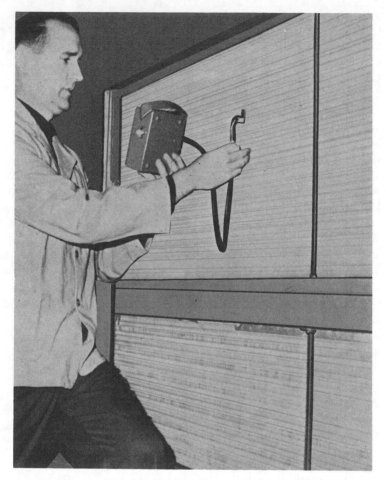

Figure 30. Measuring average velocity across a coil with an Alnor Velometer.

In Figure 31, the technician is shown using the same instrument to measure the air volume output at a diffuser. The Alnor Velometer is the recommended instrument for this purpose since most manufacturers list their grille K factors for this instrument, but other instruments can be used. The rotating vane anemometer is not well suited for this kind of measurement.

Air distribution devices of this type cannot be measured without a K factor or flow factor, as it is sometimes called because the manufacturer must test each outlet along with a particular instrument and designate the precise points on the diffuser where the instrument probe must be placed. For these devices the formula is:

$$cfm = fpm \times K \text{ factor} \qquad (29)$$

The test and balance technician must select the K factor for each type and size diffuser from the manufacturer's specification sheet, and he must use the designated instrument corresponding to the appropriate K factor for that supply outlet.

FLOW MEASURING CONES

The flow measuring hood, or cone, may be a useful device in some circumstances. By constructing a cone in accordance with Figure 32 the technician may take ceiling diffuser measurements directly without the use of a ladder and eliminate a second person. The measuring cone is particularly useful where a large number of grilles of common size are to be measured.

When flow factors for a given outlet are not available the cone may prove very helpful. By constructing the cone with a 12 × 12 in. exit opening and using the 2220 Alnor jet fitting, the readout should be in direct flow per square

Figure 31. Using the Alnor Velometer to measure the leaving air velocity at a ceiling diffuser.

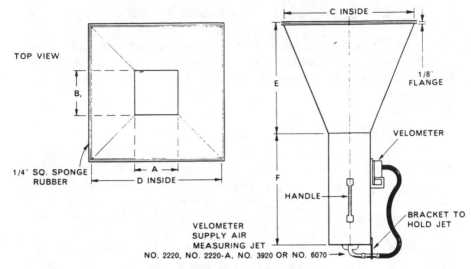

Figure 32. Flow Measuring Hood.
Dimensions:

A = 12 in.	D = 23¼ in.
B = 12 in.	E = 12 in.
C = 23¼ in.	F = 24 in.

foot. It is also possible to construct a flow measuring cone with a tight fitting exit to accommodate a rotating vane anemometer which can then be calculated in square feet of opening.

Measurement with the balancing hood or cone should not be made without calibrating the total instrument against a measured Pitot tube reading to establish correction factors for the cone with anemometer. When homemade, the cones should be fabricated from aluminum or light gage tempered wood to keep the weight down and facilitate handling the apparatus.

Aluminum, plastic, and nylon balancing cones, with velometer are available from various manufacturers in a variety of sizes for different duties.

THE PITOT TUBE AND THE BALANCING UNIT

Figure 33 shows the technician reading velocity pressure with a marked Pitot tube and manometer. The Pitot tube has been marked off in black tape to facilitate locating the traverse points. In this case the probe holes have been provided; the technician does not need to drill holes.

Pressure taps may be built-in to low pressure balancing units or into high velocity terminal units so as to make possible manometer connections for differential pressure readings without the use of the Pitot tube. Balancing units are factory built pressure recording stations placed into position during field erection of the ductwork. Such units usually have an orifice plate with a high pressure tap on the upstream side and a low pressure tap on the downstream side. When connected to a manometer, instant direct velocity readings may be made.

A high velocity terminal unit is shown in Figure 34 with an inclined manometer hook-up for differential pressure reading. Manufacturers of such terminal devices usually provide detailed testing and adjusting instructions for their products as well as calibration charts to reference the measurements.

ACCURACY IN TEMPERATURE MEASUREMENTS

Perhaps one of the weakest areas in test and balance technique is inaccurate temperature measurements—particularly the wet bulb—for total heat transfer calculations. To avoid serious errors in wet bulb temperatures, the thermometer must be at least 15 in. long, calibrated in tenths of a degree, with

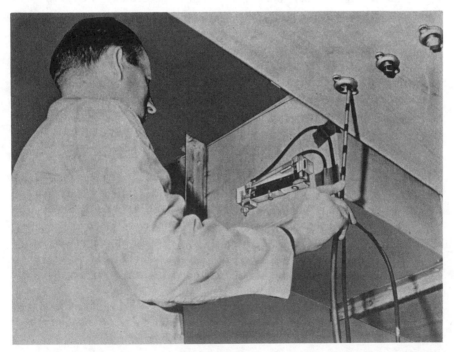

Figure 33. Using the Pitot tube with draft gage connection to measure duct velocity.

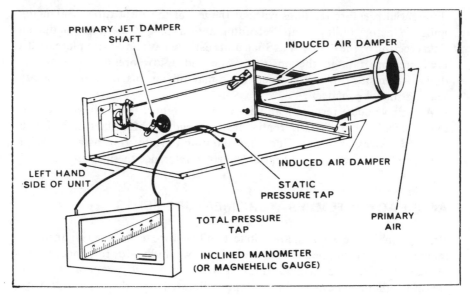

Figure 34. A high velocity terminal unit with inclined manometer hook-up for balancing.

a perfectly clean wick soaked in clean distilled water, and exposed to an air stream of about 1000 to 1500 fpm. Often, it is necessary to traverse the air pattern to collect an accurate reading in a mixed-air stream.

Familiarity with the psychrometric chart will show that small measurement errors in wet bulb temperature, say 2 or 3 degrees F, may be greatly magnified at the enthalpy scale resulting in serious miscalculation of the total heat exchange performance of a given cooling coil. For example:

An air handler is designed to deliver 5325 cfm of air and a total of 259,000 Btuh of cooling. The design mixed air entering temperature is 82.56 F DB and 67.5 F WB, and the design leaving air temperature is 55 F DB and 52 F WB. The supply duct is 30 by 30 in., and the return duct is 36 by 52 in.

The assignment is to carry out a performance test on the coil and report back all data to the design engineer. It is assumed that outside and total air have been set at the exact requirements and that the design load has been imposed on the equipment.

A nonprofessional unsheaths his standard 5 in. pocket thermometer (2 degree gradations), slides a dirty wick over the bulb, wets it thoroughly in his mouth, and inserts it into holes he has drilled in the center of the duct. He conveniently reads the entering and leaving wet bulb temperatures as 67.5 and 52 F. The enthalpy difference is 10.8, which comes to 259,000 Btuh, or 21.5 tons. The system is performing perfectly.

A professional testing and balancing engineer removes an 18 in. calibrated thermometer from his instrument case, wets the clean new wick from a small bottle of distilled water, and takes a 16 point traverse reading at both supply and discharge positions. His thermometer is divided into tenths of a degree, and the wet bulb temperature readings averge out to 64.5 F entering and 55.4 F leaving. The enthalpy difference is 6.05, which comes to 145,000 Btuh, or 12.1 tons. The equipment is performing 56% below capacity!

In this example, the difference between the wet bulb temperature of the entering mixture as measured by the nonprofessional and the professional is 3 F. Under field conditions it is not uncommon for errors of 8 F to be reported in such cases. If the configuration of the ductwork is such as to preclude a smooth mixture of outside and return air before it reaches the filter section, it

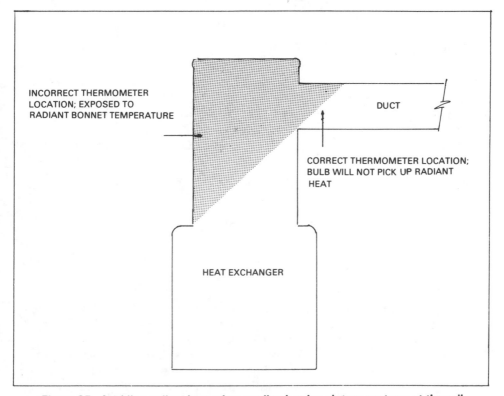

Figure 35. Avoiding radiant heat when reading leaving air temperatures at the coil.

is impossible to collect an accurate temperature reading without a traverse of the air pattern.[1]

Care must also be taken to avoid radiation effect when taking dry bulb temperature readings—particularly on the heating cycle. Figure 35 illustrates the correct location for measuring the air-temperature rise in a warm air furnace. The leaving bonnet temperature should be read as close to the heat exchanger as possible *without being exposed to the radiation effect.*

[1] J. Gladstone "Testing and Balancing of Air Conditioning Systems: State of the Arts," *Heating Piping & Air Conditioning*, Feb. 1968.

8

Duct Pressures and How They Are Measured

When air travels at a specific velocity it creates a pressure corresponding to the velocity of flow; this is a measure of the kinetic energy in the fluid and it is known as the *velocity pressure* (Vp). Velocity pressure is always exerted in the direction of air flow. Assuming gravitational acceleration (g) equal to 32.174 ft/sec^2, and the density of air equal to 0.075 lb/ft^3, the relationship between the velocity and the velocity pressure may be expressed by the following formulas;

$$Vp = \left(\frac{V}{4005}\right)^2 \tag{30}$$

$$V = 4005\sqrt{Vp}$$

It is therefore a simple matter to determine the velocity (fpm) of air stream if the VP can be measured. For example, a Pitot tube manometer hook-up reads 0.250 in. water, substituting for the above equation;

$$4005\sqrt{0.250} = 2002 \text{ fpm}$$

Table 10 gives the calculated conversions of velocity in fpm from velocity pressure Vp. This table is an important tool in test and balance work; it covers the complete range from 300 to 5250 fpm.

Independent of its velocity, air, when confined within an enclosure such as a duct or tank will exert itself perpendicularly to the walls of the enclosure. This is the compressive pressure existing in a fluid, and it is known as the static pressure (Sp). Unlike velocity pressure, which is always positive, static

Table 10. Conversion of Velocity Pressures (VP) to Velocity (FPM).

VELOCITY PRESSURE IN. OF H_2O	VELOCITY FPM	VELOCITY PRESSURE IN. OF H_2O	VELOCITY FPM	VELOCITY PRESSURE IN. OF H_2O	VELOCITY FPM
0.0056	300	0.100	1300	0.681	3300
0.0060	310	0.105	1350	0.695	3350
0.0064	320	0.122	1400	0.722	3400
0.0068	330	0.131	1450	0.740	3450
0.0072	340	0.140	1500	0.766	3500
0.0076	350	0.149	1550	0.785	3550
0.0081	360	0.160	1600	0.810	3600
0.0086	370	0.169	1650	0.825	3650
0.0090	380	0.181	1700	0.860	3700
0.0095	390	0.190	1750	0.875	3750
0.010	400	0.208	1800	0.90	3800
0.011	425	0.213	1850	0.92	3850
0.012	450	0.225	1900	0.95	3900
0.014	475	0.237	1950	0.97	3950
0.016	500	0.250	2000	1.00	4000
0.017	525	0.262	2050	1.02	4050
0.019	550	0.276	2100	1.04	4100
0.021	575	0.290	2150	1.06	4150
0.022	600	0.302	2200	1.08	4200
0.024	625	0.319	2250	1.12	4250
0.026	650	0.331	2300	1.14	4300
0.028	675	0.348	2350	1.17	4350
0.030	700	0.360	2400	1.20	4400
0.033	725	0.375	2450	1.23	4450
0.035	750	0.391	2500	1.25	4500
0.038	775	0.408	2550	1.28	4550
0.040	800	0.422	2600	1.31	4600
0.042	825	0.439	2650	1.34	4650
0.045	850	0.456	2700	1.37	4700
0.048	875	0.467	2750	1.40	4750
0.050	900	0.490	2800	1.43	4800
0.053	925	0.509	2850	1.46	4850
0.057	950	0.530	2900	1.50	4900
0.059	975	0.543	2950	1.53	4950
0.062	1000	0.562	3000	1.56	5000
0.066	1050	0.576	3050	1.59	5050
0.075	1100	0.600	3100	1.62	5100
0.082	1150	0.615	3150	1.65	5150
0.090	1200	0.640	3200	1.68	5200
0.097	1250	0.665	3250	1.72	5250

pressure, when it is above atmospheric pressure, will be positive but when below atmospheric pressure will be negative. The discharge side of a fan in a supply system will read a positive pressure; the inlet side of the fan in an exhaust system will read a negative or minus pressure.

Static pressure is exerted whether air is at rest or in motion. The algebraic sum of static pressure and velocity pressure gives the *total pressure* (TP), therefore;

$$Vp = Tp - Sp \qquad (31)$$

Demonstrating the performance of pressures encountered under actual conditions in a ventilation system, assume a length of duct capped at both ends and sealed tight. This rectangular chamber contains a mass of air at 0.06 psi. If a Pitot tube is inserted through a test tap, the static pressure on an inclined manometer would read 1.66 in. of water (See conversion Table 32). Since the air is at rest and no velocity pressure is exerted, the Vp reading on the manometer would be 0 in. of water and the Tp reading would be 1.66 in. as shown in Figure 36.

Now assume a fan is connected to one end of the duct and air is pumped through the duct at 4900 fpm (1.5 in. of water Vp). As seen in Figure 37 the air velocity at 4900 fpm would convert to 1.5 in. water velocity pressure. Subtracting the velocity pressure from the total pressure.

$$1.66\ Tp - 1.5\ Vp = 0.16\ in.\ Sp$$

The manometer does not sense the actual velocity pressure directly but by using the Pitot tube hook-up with the static opening connected to the low pressure side of the gage, and the total pressure opening connected to the high pressure side of the gage, the manometer will read the difference between the two, or the *velocity pressure.*

Velocity pressure and static pressure changes in the duct work with every change in the duct configuration, but the total pressure on the other hand, remains constant. Hence, as the velocity pressure decreases the static pressure increases and vice versa, because the static pressure is always the difference between the total pressure and the velocity pressure. It should be remembered, however, that in an actual duct system, the internal friction will cause a loss of total pressure.

The static pressure in an exhaust system is always below atmospheric pressure, and it is customary among ventilation engineers to omit the minus sign affecting the static (gage pressure). These engineers know, of course, that the total pressure is higher than the static pressure by the amount of the velocity pressure. But total pressure is seldom considered in the field of

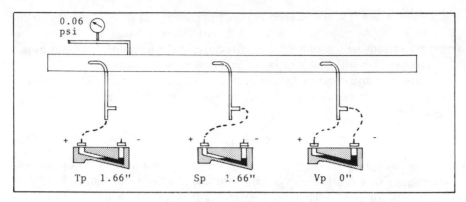

Figure 36. Pressure At Rest.

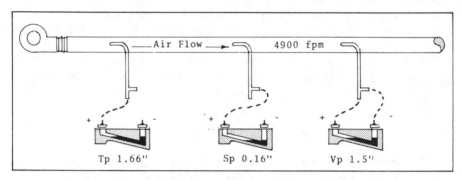

Figure 37. Pressure Above Atmosphere.

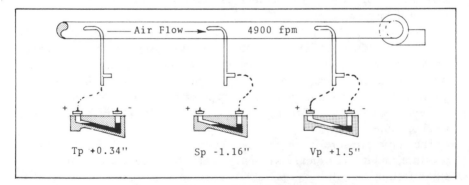

Figure 38. Pressure Below Atmosphere.

ventilation; all computation for design and testing being based on static pressure.

In Figure 38 we assume the same condition as in Figure 37, but now the fan is on the suction side. The pressure would then be below atmospheric and the manometer readings could be; velocity pressure + 1.50 in., static pressure − 1.16 in.; and total pressure, + 0.34 in.

9

Air Mixtures At The Coil

Determining the correct proportion of outside to total air is basic to the proper balancing of any system. Because of close connections between the outside air duct and the equipment, the Pitot tube traverse method of measuring outside air quantities is either extremely difficult or altogether impossible. Therefore, the most practical method of setting outside air proportional dampers is the *mixed air temperature* method.

The *mixed air temperature* method is dependent on the formula;

$$T_{mix} = (\% \, OA \times T_{oa}) + (\% \, RA \times T_{ra}) \tag{32}$$

where

T_{mix} = Temperature of air mixture
T_{oa} = Temperature of outdoor air
T_{ra} = Temperature of return air
%OA = Percentage of outdoor air to total air
%RA = Percentage of return air to total air

The above formula may be restated as;

$$\%OA = \frac{T_{ra} - T_{mix}}{T_{ra} - T_{oa}} \times 100$$

Example: A system is designed for 15,000 cfm total air. The minimum setting for the outside air damper calls for 2250 cfm. Determine the correct setting of the outdoor air damper.

Step 1. Using the Pitot tube traverse at the proper point of the supply duct, set the fan speed to deliver 15,000 cfm total air.

Step 2. Insert a set of calibrated thermometers at strategic points to measure the temperature of the outdoor air, return air and mixed air. Assume that these read 91 F outside, 75 F return, and 76 F mixture.

Step 3. Find the design percentage of outdoor air:

$$\frac{2250}{15,000} = 0.15$$

Step 4. Calculate the air mixture temperature required to give 15% outdoor air by using the above formula.

$$T_{mix} = (0.15 \times 91) + (0.85 \times 75) = 77.4 \text{ F}$$

Step 5. Since the recorded mixture temperature was 76 F and the required mixture temperature is 77.4 F, the system is obviously short of outdoor air (by formula, 6%); therefore, open the damper slightly to allow more outdoor air until the air mixture temperature rises to 77.4 F.

Example: Outdoor air at 91 F db and 79 F wb is to be mixed with return air at 78 F db and 55% rh. Find the percentage of OA that will result in a 70 F wb mixture.

Solution: Plot the outside air and room air on the psychrometric chart and connect the slope. Now find 70 F wb on the saturation curve and follow its line to the point where it intersects the outside air/room air slope. From the point of intersection, drop down to the dry bulb scale and read 81.25 F db. Now, transpose Equation 32 to;

$$\% \text{ OA} = \frac{T_{ra} - T_{mix}}{T_{ra} - T_{oa}} \times 100$$

$$= \frac{(81.25 - 78)(100)}{(91 - 78)} = \frac{325}{13} = 25\% \text{ OA}$$

The mixture conditions for any number of air streams may be found using Equation 32.

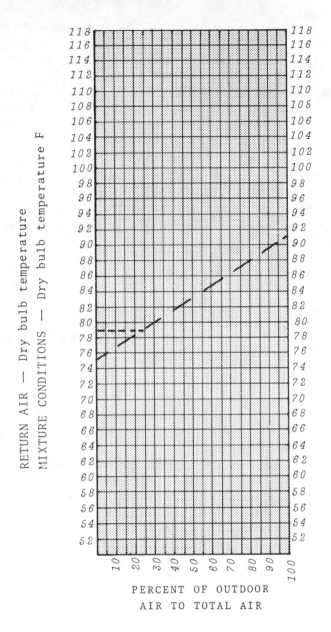

Example: Outdoor air reads 91 F, return air reads 75 F and mixture conditions read 79 F. Find the percent of outdoor air. Lay a straight edge across 91 F OA and 75 F RA. Enter left ordinate at 79 F mix and move to the right to point of intersection; read 25% on bottom scale.

Figure 39. Speed-O-Graph. Mixed Air Temperature Method Of Adjusting Dampers For Correct Percentages Of Outdoor Air.

Example:

Conditions for a multizone unit are as follows;
   RA zone 1 = 1000 cfm at 78 F
   RA zone 2 = 1500 cfm at 76 F
   RA zone 3 = 2000 cfm at 77 F
          OA =   500 cfm at 91 F
Find the temperature of the air mixture entering the coil.
Solution:
   1000 cfm + 1500 cfm + 2000 cfm + 500 cfm = 5000 cfm, total air
   1000/5000 = 20%
   2000/5000 = 40%
   1500/5000 = 30%
    500/5000 = 10%

$(0.20 \times 78) + (0.40 \times 76) + (0.30 \times 77) + (0.10 \times 91) =$
   15.6   +   30.4   +   23.1   +   9.1   = 78.2 F

The resulting temperature of the dry bulb mixture at the coil will be 78.2 F.

   Figure 39 offers a Speed-O-Graph for determining outside air quantities using the mixed air temperature method.

10

Basic Formulas For Pumps

TO FIND THE BRAKE HORSEPOWER (BHP)

$$bhp = \frac{gpm \times tdh \times G}{3960 \times efficiency} \qquad (33)$$

where
  tdh = total dynamic head, feet
   G = specific gravity

The pump efficiency must be provided by the manufacturer. Where the efficiency is not known, a good rule-of-thumb is to allow 70% as the efficiency factor. The specific gravity for cold water is taken as 1. Therefore, Equation 33 simplified becomes;

$$bhp = \frac{gpm \times tdh}{2800}$$

Example: A pump delivers 300 gpm against 102 ft head, the pump efficiency is unknown.
  Find: The brake horsepower
  Solution:

$$\frac{300 \times 102}{2800} = 10.9 \text{ bhp}$$

To Find the Quantity of Water Pumped Where the Horsepower Is Known

$$gpm = \frac{bhp \times 2800}{tdh}$$

Example: A 15 hp pump is moving water against a head of 150 ft.
  Find: The quantity of water
  Solution:

$$Q = \frac{15 \times 2800}{150} = 280 \text{ gpm}$$

As The Specific Gravity Varies

For a given head in feet the hp will vary according to the specific gravity of the liquid. The specific gravity of gasoline, for example, is 0.75 and the specific gravity of brine is 1.2; therefore, for 200 gpm at a total head of 100 ft and a pump efficiency of 70%.

$$(\text{water}) \text{ hp} = \frac{200 \times 100 \times 1.0}{3960 \times 0.70} = 7.2$$

$$(\text{gasoline}) \text{ hp} = \frac{200 \times 100 \times 0.75}{3960 \times 0.70} = 5.4$$

$$(\text{brine}) \text{ hp} = \frac{200 \times 100 \times 1.2}{3960 \times 0.70} = 8.7$$

$$(210 \text{ F water}) \text{ hp} = \frac{200 \times 100 \times 0.960}{3960 \times 0.70} = 6.9$$

Centrifugal pump performance can be expressed by the following pump laws:
  a. Capacity varies directly as the impeller diameter
  b. Head varies directly as the square of the diameter
  c. Power varies directly as the cube of the diameter.

HYDRAULIC DEFINITIONS

PRESSURE DROP is the loss in pressure as liquid flows through a unit of resistance such as a condenser, control valve, or tower.
FRICTION LOSS is the loss in pressure owing to the resistance of pipe and fittings measured in equivalent length of pipe in feet.
STATIC SUCTION LIFT is the vertical distance in feet from the centerline of the pump to the free level of the liquid to be pumped. Liquid source is below the pump.
TOTAL DYNAMIC SUCTION LIFT is the *static suction lift* plus the *friction losses, pressure drop,* and *velocity head.*
STATIC SUCTION HEAD is the vertical distance in feet from the center line

of the pump to the free level of the liquid to be pumped. Source of supply is above the pump.

TOTAL DYNAMIC SUCTION HEAD is the *static suction head* minus the *friction losses, pressure drop,* and *velocity head* in the suction line.

TOTAL STATIC HEAD is the vertical distance in feet between the free level of the source of supply and the point of free discharge or to the level of the free surface of the discharge water.

TOTAL DYNAMIC HEAD is the *total static head* plus all *friction losses and pressure drops* through the entire piping system, plus the *velocity head.*

VELOCITY HEAD is the equivalent head in feet through which the liquid must fall to attain the same velocity, or the head necessary to accelerate the liquid.

Velocity head formula:

$$H_v = \frac{V^2}{2g}$$

where

H_v = velocity head, ft

V = velocity of liquid through the pipe in fps

g = acceleration due to gravity, 32.2 fps

Velocity Head—Velocity Conversion

Velocity, fps	2	3	4	5	6	7	8	9	10	11	12
Velocity head, ft.	0.06	0.14	0.25	0.39	0.56	0.76	1.01	1.25	1.55	1.87	2.24

11

Heat Transfer

Because all environmental systems are based on controlling the transfer of heat energy, familiarity with the basic heat transfer theory is requisite for the test and balance technician.

Heat transfer takes place by 1.) *Radiation*: the transmission of heat energy from one body to another by passing through a medium which is transparent to it. It may be reflected and refracted. 2.) *Convection*: the heat transmitted from one object to another through a medium of moving gas or liquid. 3.) *Conduction*: the transmission of heat between two particles that are in contact with one another.

As a thermodynamic process heat transfer is a function of weight, specific heat and temperature difference. It is expressed mathematically as:

Heat = weight of the transfer medium × specific heat × temperature difference

$$Btu = W \times C \times \Delta t \tag{34}$$

where

W = weight in lbs

C = specific heat in Btu/lb/degree F

Δt = temperature difference in degrees F

Specific heat (C) is a reference index. It is used to measure the amount of heat necessary to raise the temperature of one pound of any substance one degree F. Different substances have different specific heats; specific heat changes with temperature pressure change. Table 11 gives specific heats for various elements. For the test and balance technician it is important to

Table 11. Specific Heat Of Various Substances.

LIQUIDS

Material	Temp. °F. Range	Specific Heat
Alcohol, Ethyl	32.0	0.548
Alcohol, Methyl	5.0	0.59
Anilin	60.0	0.514
Benzol	105.0	0.423
CaCl2Sp.Gr. 1.14	5.0	0.764
CaCl2Sp.Gr. 1.20	-4.0	0.695
CaCl2Sp.Gr. 1.26	-4.0	0.651
Ethyl Ether	32.0	0.529
Glycerine	59 – 120	0.576
NaCl plus 10 H2O	64.0	0.791
NaCl plus 200 H2O	64.0	0.978
Napthalene	185.0	0.396
Nitro Benzole	84.0	0.362
Oils: Castor	44.0	0.434
Olive		0.471
Sesame		0.387
Turpentine	32.0	0.411
Petroleum	70 – 135	0.511
Sea Water Sp.Gr.1.0235	64.0	0.938
Toluol	150.0	0.490

GASES

Material	Temp. °F. Range	Sp.Ht.at Constant Pressure	Sp.Ht.at Constant Volume
Acetone	79 – 230	0.3468	
Air	-22 – 50	0.2377	0.168
Air	32 – 400	0.2375	
Alcohol, C_2H_5OH	110 – 400	0.4534	0.399
Alcohol, CH_3OH	110 – 400	0.4580	
Ammonia	73 – 212	0.5202	0.299
Benzene C_6H_6	94 – 235	0.2990	
Carbon Dioxide CO_2	-20 – 45	0.1843	0.171
Carbon Monoxide CO	74 – 210	0.2425	0.176
Chlorine	61 – 700	0.1125	
Chloroform $CHCl_3$	80 – 230	0.1441	
Ether $C_4H_{10}O$	77 – 232	0.4280	
Hydrochloric Acid	50 – 212	0.1910	
Hydrogen	54 – 300	3.4090	2.412
Methane CH_4	66 – 390	0.5929	
Nitrogen	32 – 390	0.2438	
Nitrous Oxide	80 – 220	0.2126	
Oxygen	50 – 400	0.2175	0.155
Sulfur Dioxide	32	0.1544	
Water Vapor	32	0.4655	
Water Vapor	212	0.421	0.346

SOLIDS

Material	LB/CU.FT.	Specific Heat	Material	LB/CU.FT.	Specific Heat
Asbestos	43.0	0.20	Cotton, Loose	30.0	0.32
Ashes	43.0	0.20	Fats	58.0	0.46
Bakelite, Laminated	86.0	0.35	Glass, Common	164.0	0.199
Benzol	55.0	0.42	Glass, Plate	161.0	0.161
Borax		0.24	Graphite	126.0	0.201
Bronze, Phosphor	554.0	0.09	Gypsum, Loose	70.0	0.26
Calcium Carbonate	177.0	0.18	Ice $-14°$	57.5	0.53
Calcium Sulfate	185.0	0.27	Litharge		0.21
Carborundum	195.0	0.16	Mica		0.10
Cellulose	94.0	0.37	Paper	58.0	0.324
Celluloid	90.0	0.36	Paraffin, $4°$ to $40°$		0.377
Chalk	142.0	0.21	Paraffin, $32°$ to $68°$		0.694
Coke	75.0	0.20	Plaster Paris	103.0	1.14
Concrete, Stone	147.0	0.19	Rubber	59.0	0.48
Concrete, Cinder	105.0	0.18	Sugar	100.0	0.22
Cork	15.0	0.48	Sulfur	126.0	0.17
Corundum	247.0	0.20	Wood	44.0	0.373
Cotton, Baled	93.0	0.32			

remember that the specific heat of standard air is 0.24 (1005 J/kg/K), and the specific heat of standard water is 1.0 (4190 J /kg/K).

STANDARD AIR

Standard air is air at 70 F (21 C) and 29.92 in. of mercury (101 kPa), or 14.7 psig (zero psia). The density—or weight—of standard air is 0.075 pounds per cubic foot (1.2 kg/m³). The specific volume is the reciprocal of the density; therefore, for standard air the volume is 1/0.075 = 13.33 cubic feet per pound (0.833 m³/kg). Whenever either density or volume is known, the other variable may be found by taking the reciprocal; 1/13.33 = 0.075.

For every 10% increase in relative humidity the dry bulb temperature can be reduced about 1 degree F.

HEAT TRANSFER FORMULAS

Equation 34 is the basis of all heat transfer calculations for convective and conductive flow. Radiant heat calculations are made from the Stefan-Bolzmann law which states that the radiation from a black body to another body is proportional to the difference between the fourth power of the absolute temperatures of the two bodies:

$$Q = \frac{0.174 \ A \ E \ (T_2^4 - T_1^4)}{10^8} \tag{35}$$

where
  Q = radiation in Btuh
  A = area in ft^2
  E = emissivity of radiating body
  T_2 = absolute temperature of radiating body
  T_1 = absolute temperature of receiving body
Radiant heat calculations are seldom required for testing and balancing environmental systems. When required, tables for this formula may be found in handbooks.
For overall heat transmission the working formula is;

$$Btuh = U \times A \times \Delta t \tag{36}$$

where
  U = overall coefficient, Btu/(hr) (F) (ft^2)
  A = area in ft^2
  Δt = temperature difference between air on two sides of a structure.

How does this square with Equation 34—the basis of all heat transfer? The coefficient U must be taken from tables which are, in effect, precalculated factors of the weight and specific heat per square foot of various materials, or the reciprocal of the resistance factors for each substance. That is, the coefficient U is a convenient tabular solution whereby the weight × the specific heat (W × C) has been converted into a factor per square foot value, thus simplifying the calculations for walls, floors, ceilings, etc.

In the same manner, the formulas for standard air and standard water have been simplified for more convenient use. Substituting for standard air in Equation 34.

$$Btu = W \times C \, \Delta t$$

$$= 0.075 \times 0.24 \times \Delta t$$

Unfortunately, the industry adopted the use of cfm years ago and it now becomes necessary to convert minutes to solve for Btu/hour:

$$0.075 \times 0.24 \times 60 \text{ min} \times cfm \times \Delta t = Btuh$$

$$0.075 \times 0.24 \times 60 = 1.08$$

Air Sensible Heat Transfer, simplified, becomes:

$$Btuh \text{ (sensible)} = cfm \times 1.08 \times \Delta t \tag{37}$$

$$\frac{Btuh}{1.08 \times \Delta t} = cfm$$

This is the formula for sensible heat transfer of air only. It cannot be used for latent heat calculations on cooling coils.
To convert cfm to lb/hr,

$$lb \text{ of air per hour} = cfm \times 4.5 \tag{38}$$

$$\frac{lb/hr}{4.5} = cfm$$

where

$$4.5 = \frac{60 \text{ min}}{1 \text{ hr}} \times 0.075$$

Air Latent Heat Transfer, simplified, becomes,

$$Btuh \text{ (latent)} = cfm \times 0.68 \times \Delta gr/lb \tag{39}$$

The derivation of 0.68 is $\dfrac{60}{13.33} \times \dfrac{1076}{7000} = 0.68$

where

$\quad$ 60 = min/hr

$\quad$ 13.33 = specific volume of standard air

$\quad$ 1076 = average heat of vaporization

$\quad$ 7000 = grains per pound

and the difference in grains per pound is taken from the psychrometric chart.
Air Total Heat Transfer, simplified, becomes,

$$\text{Btuh (total)} = \text{cfm} \times 4.5 \times \Delta h \qquad (40)$$

where

$\quad$ 4.5 = 60 min $\times$ 0.075 lb/ft.^{3}

$\quad \Delta h$ = the enthalpy difference in Btu per lb as read from the psychrometric chart

Water is usually measured in gallons per minute. One gallon of standard water weights 8.33 lb. The specific heat of water is 1. Therefore, substituting for Equation 34.

$$\text{Btu} = \text{W} \times \text{C} \times \Delta t$$
$$8.33 \times 1 \times 60 \text{ min} \times \text{gpm} \times \Delta t = \text{Btu}$$

Water Heat Transfer, simplified, becomes:

$$\text{Btuh} = 500 \times \text{gpm} \times \Delta t \qquad (41)$$

$$\text{gpm} = \dfrac{\text{Btuh}}{500 \times \Delta t}$$

Example:

$\quad$ *Given:* A condenser is operating at 100 gpm. Water on at 60 F, and off at 70 F. The condensing temperature is 80 F and the condensing area is 300 ft^2.

$\quad$ *Find:* $\quad$ A. Btu/min

$\qquad\qquad$ B. Arithmetical mean temperature difference (AMTD)

$\qquad\qquad$ C. Rate of heat transfer, Btu/ft^2/hr/F

$\quad$ *Solution:* $\quad$ A. (500) (70–60) (100) = 500,000 Btuh

$\qquad\qquad\quad$ B. 80 F − 60 F = 20 F (water on)

$\qquad\qquad\qquad\quad$ 80 F − 70 F = 10 F (water off)

$\qquad\qquad\qquad\quad$ 20 F + 10 F = 30 F 30/2 = 15 F AMTD

C. Rate of heat transfer $= \dfrac{\text{Btuh}}{\text{ft}^2 \times \Delta t}$

$\qquad\qquad\qquad = \dfrac{500{,}000 \text{ Btuh}}{300 \text{ ft.}^2 \times 15 \text{ AMTD}}$

$\qquad\qquad\qquad = 111 \text{ Btu/ft}^2/\text{hr}/\text{F}$

LOG MEAN TEMPERATURE DIFFERENCE (LMTD)

Where two fluids are used in a heat transfer process the temperature difference at the end of the process will be less than at the beginning, thus a square foot of exchange surface at the end will do less work than an equal area at the beginning. The heat exchange will follow a logarithmic curve. The log mean temperature difference will give correct values for calculating heat transfer rates in double pipe energy recovery systems:

$$\text{LMTD} = \frac{\Delta t_L - \Delta t_s}{\log_e (\Delta t_L / \Delta t_s)} \qquad (42)$$

where

Δt_L = the larger temperature difference

Δt_s = the smaller temperature difference

$\log_e$ = Naperian logarithm

Example:

Given: A fluid enters a heat exchanger at 80 F and leaves at 46 F. Water

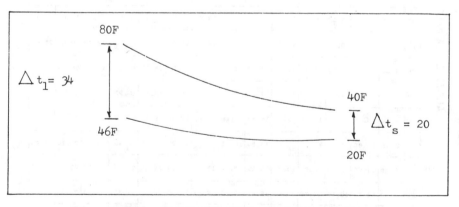

Figure 40. Mean Temperature Difference In Heat Exchanger.

Table 12. Logarithmic Mean Temperature Difference, LMTD.

LARGER OR SMALLER TEMPERATURE DIFFERENCE; ENTER EITHER SIDE

LARGER OR SMALLER TEMPERATURE DIFFERENCE; ENTER EITHER SIDE

	1	2	3	4	5	6	7	8	9	10	12	14	16	18
1	1.00	1.40	1.51	2.05	2.45	2.75	3.00	3.25	3.51	3.80	4.40	4.90	5.40	5.80
2	1.40	2.00	2.45	2.80	3.25	3.60	4.00	4.35	4.70	5.00	5.60	6.20	6.75	7.25
3	1.51	2.45	3.00	3.45	3.80	4.25	4.70	5.10	5.50	5.80	6.50	7.20	7.80	8.45
4	2.05	2.80	3.45	4.00	4.45	4.90	5.35	5.80	6.20	6.60	7.30	8.00	8.65	9.25
5	2.45	3.25	3.80	4.45	5.00	5.45	5.95	6.40	6.75	7.20	8.00	8.75	9.50	10.20
6	2.75	3.60	4.25	4.90	5.45	6.00	6.50	7.00	7.50	7.90	8.75	9.50	10.20	11.00
7	3.00	4.00	4.70	5.35	5.95	6.50	7.00	7.50	8.00	8.45	9.30	10.20	10.90	11.70
8	3.25	4.35	5.10	5.80	6.40	7.00	7.50	8.00	8.50	9.00	10.00	10.80	11.65	12.50
9	3.51	4.70	5.50	6.20	6.75	7.50	8.00	8.50	9.00	9.50	10.50	11.45	12.25	13.00
10	3.80	5.00	5.80	6.60	7.20	7.90	8.45	9.00	9.50	10.00	11.00	12.00	12.90	13.70
12	4.40	5.60	6.50	7.30	8.00	8.75	9.30	10.00	10.50	11.00	12.00	13.00	14.00	14.85
14	4.90	6.20	7.20	8.00	8.75	9.50	10.20	10.80	11.45	12.00	13.00	14.00	15.00	16.00
16	5.40	6.75	7.80	8.65	9.50	10.20	10.90	11.65	12.25	12.90	14.00	15.00	16.00	17.00
18	5.80	7.25	8.45	9.25	10.20	11.00	11.70	12.50	13.00	13.70	14.85	16.00	17.00	18.00
20	6.20	7.80	9.00	9.80	10.80	11.60	12.45	13.20	13.80	14.50	15.70	16.80	18.00	19.00
22	6.70	8.30	9.50	10.50	11.45	12.25	13.20	13.90	14.20	15.25	16.60	17.80	19.00	20.10
24	7.10	8.80	10.00	11.00	12.00	12.95	13.75	14.55	15.25	16.00	17.45	18.70	19.80	21.00
26	7.50	9.25	10.55	11.60	12.65	13.60	14.50	15.25	16.00	16.75	18.20	19.50	20.70	22.00
28	8.00	9.80	11.10	12.20	13.25	14.20	15.20	16.00	16.75	17.50	18.95	20.30	21.50	22.80
30	8.40	10.30	11.60	12.75	13.90	14.90	15.80	16.65	17.50	18.20	19.70	21.20	22.40	23.70
32	8.75	10.80	12.20	13.30	14.50	15.50	16.45	17.25	18.20	18.80	20.45	21.85	23.10	24.50
34	9.20	11.30	12.70	13.85	15.00	16.10	17.00	17.95	18.80	19.50	21.20	22.70	23.90	25.20
36	9.60	11.80	13.25	14.50	15.60	16.65	17.70	18.55	19.50	20.25	21.85	23.45	24.70	26.00
38	10.00	12.30	13.75	15.00	16.20	17.25	18.40	19.22	20.20	21.00	22.60	24.15	25.50	26.70

20	22	24	26	28	30	32	34	36	38	40	42	44
6.20	6.70	7.10	7.50	8.00	8.40	8.75	9.20	9.60	10.00	10.40	10.80	11.25
7.80	8.30	8.80	9.25	9.80	10.30	10.80	11.30	11.80	12.30	12.75	13.20	13.70
9.00	9.50	10.00	10.55	11.10	11.60	12.20	12.70	13.25	13.75	14.30	14.80	15.40
9.80	10.50	11.00	11.60	12.20	12.75	13.30	13.85	14.50	15.00	15.60	16.20	16.70
10.80	11.45	12.00	12.65	13.25	13.90	14.50	15.00	15.60	16.20	16.75	17.40	17.90
11.60	12.25	12.95	13.60	14.20	14.90	15.50	16.10	16.65	17.25	17.80	18.50	19.05
12.45	13.20	13.75	14.50	15.20	15.80	16.45	17.00	17.70	18.40	19.00	19.50	20.20
13.20	13.90	14.55	15.25	16.00	16.65	17.25	17.95	18.55	19.22	19.80	20.50	21.20
13.80	14.20	15.25	16.00	16.75	17.50	18.20	18.80	19.50	20.20	20.85	21.50	22.10
14.50	15.25	16.00	16.75	17.50	18.20	18.80	19.50	20.25	21.00	21.60	22.25	22.85
15.70	16.60	17.45	18.20	18.95	19.70	20.45	21.20	21.85	22.60	23.45	24.00	24.75
16.90	17.80	18.70	19.50	20.30	21.20	21.85	22.70	23.45	24.15	24.90	25.60	26.30
18.00	19.00	19.80	20.70	21.50	22.40	23.10	23.90	24.70	25.50	26.40	27.00	27.75
19.00	20.10	21.00	22.00	22.80	23.70	24.50	25.20	26.00	26.70	27.65	28.40	29.10
20.00	21.10	22.10	23.00	24.00	24.80	25.70	26.50	27.30	28.10	29.00	29.75	30.60
21.10	22.00	23.10	24.00	24.90	25.85	26.75	27.60	28.50	29.40	30.20	31.10	31.86
22.10	23.10	24.00	25.00	26.00	27.00	27.90	28.80	29.70	30.50	31.40	32.30	33.20
23.00	24.00	25.00	26.00	27.00	28.00	29.00	29.90	30.75	31.60	32.50	33.45	34.30
24.00	24.90	26.00	27.00	28.00	29.00	30.00	30.95	31.90	32.80	33.70	34.55	35.45
24.80	25.85	27.00	28.00	29.00	30.00	31.00	32.00	33.00	34.00	34.85	35.75	36.65
25.70	26.75	27.90	29.00	30.00	31.00	32.00	33.00	34.00	35.05	36.00	36.90	37.80
26.50	27.60	28.80	29.90	30.95	32.00	33.00	34.00	35.00	36.10	37.10	38.00	38.90
27.30	28.50	29.70	30.75	31.90	33.00	34.00	35.00	36.00	37.00	38.05	39.10	40.00
28.10	29.40	30.50	31.60	32.80	34.00	35.05	36.10	37.00	38.00	39.00	40.10	41.10

Table 12. (continued) Logarithmic Mean Temperature Difference.

LARGER OR SMALLER TEMPERATURE DIFFERENCE; ENTER EITHER SIDE

LARGER OR SMALLER TEMPERATURE DIFFERENCE; ENTER EITHER SIDE

	1	2	3	4	5	6	7	8	9	10	12	14	16	18
40	10.40	12.75	14.30	15.60	16.75	17.80	19.00	19.80	20.85	21.60	23.45	24.90	26.40	27.65
42	10.80	13.20	14.80	16.20	17.40	18.50	19.50	20.50	21.50	22.25	24.00	25.60	27.00	28.40
44	11.25	13.70	15.40	16.70	17.90	19.05	20.20	21.20	22.10	22.85	24.75	26.30	27.75	29.10
46	11.70	14.10	15.90	17.20	18.50	19.07	20.70	21.72	22.75	23.55	25.50	27.00	28.50	29.90
48	12.00	14.60	16.45	17.80	19.00	20.20	21.30	22.40	23.40	24.20	26.20	27.75	29.25	30.70
50	12.50	15.00	16.80	18.25	19.50	20.80	21.80	23.00	24.00	24.80	26.85	28.50	30.00	31.60
52	12.90	15.40	17.30	18.80	20.00	21.35	22.45	23.50	24.55	25.50	27.50	29.15	30.70	32.20
54	13.30	15.75	17.80	19.30	20.60	21.80	23.00	24.10	25.10	26.10	28.10	29.75	31.40	32.80
56	13.70	16.20	18.30	19.80	21.20	22.45	23.50	24.70	25.70	26.70	28.70	30.45	32.00	33.60
58	14.00	16.50	18.70	20.40	21.70	22.85	24.10	25.20	26.20	27.25	29.30	31.10	32.70	34.40
60	14.40	16.90	19.20	20.90	22.20	23.45	24.65	25.75	26.75	27.80	29.90	31.72	33.40	35.00
62	14.70	17.30	19.70	21.40	22.70	23.85	25.20	26.35	27.35	28.50	30.50	32.40	34.00	35.70
64	15.10	17.70	20.20	21.90	23.10	24.45	25.70	26.90	27.85	29.00	31.10	33.00	34.70	36.50
66	15.40	18.10	20.60	22.40	23.55	24.85	26.25	27.50	28.45	29.70	31.70	33.70	35.25	37.20
68	15.75	18.50	21.10	22.80	24.00	25.45	26.75	28.00	29.00	30.25	32.25	34.30	36.00	37.90
70	16.10	18.90	21.50	23.40	24.50	25.85	27.30	28.50	29.50	30.85	32.80	35.00	36.60	38.50
72	16.45	19.25	22.00	23.80	25.00	26.45	27.80	29.00	30.10	31.45	33.50	35.60	37.25	39.20
74	16.75	19.60	22.40	24.25	25.45	26.85	28.40	29.60	30.70	32.00	34.00	36.20	37.90	39.80
76	17.10	20.00	22.80	24.75	25.80	27.40	29.00	30.15	31.20	32.50	34.60	36.80	38.50	40.50
78	17.40	20.45	23.20	25.20	26.40	27.80	29.50	30.70	31.75	33.10	35.20	37.40	39.10	41.20
80	17.70	20.80	23.70	25.70	26.75	28.40	30.00	31.15	32.40	33.70	35.70	38.00	39.70	41.75

20	22	24	26	28	30	32	34	36	38	40	42	44
29.00	30.20	31.40	32.50	33.70	34.85	36.00	37.10	38.05	39.00	40.00	41.00	42.00
29.75	31.10	32.30	33.45	34.55	35.75	36.90	38.00	39.10	40.10	41.00	42.00	43.00
30.60	31.85	33.20	34.30	35.45	36.65	37.80	38.90	40.00	41.10	42.00	43.00	44.00
31.40	32.75	34.05	35.20	36.30	37.50	38.70	39.80	40.90	42.00	43.00	44.00	44.90
32.20	33.60	34.95	36.00	37.15	38.40	39.60	40.75	41.85	43.00	43.95	44.85	46.08
33.00	34.40	35.75	36.90	38.00	39.30	40.50	41.60	42.75	43.90	44.80	46.00	47.00
33.70	35.15	36.55	37.70	38.80	40.15	41.40	42.50	43.70	44.70	45.50	47.06	48.10
34.45	35.90	37.35	38.50	39.60	41.00	42.30	43.40	44.60	45.63	46.71	47.79	49.14
35.20	36.70	38.15	39.35	40.40	41.80	43.20	44.30	45.36	46.36	47.60	48.72	49.84
35.80	37.35	38.90	40.20	41.25	42.70	44.10	45.24	46.51	47.27	48.43	49.70	50.75
36.50	38.10	39.70	41.00	42.00	43.50	45.00	46.02	47.10	48.36	49.50	50.40	51.72
37.25	38.85	40.45	41.70	42.85	44.45	45.57	46.81	47.86	49.29	50.40	51.33	52.39
38.00	39.55	41.20	42.40	43.70	45.12	46.30	47.55	48.96	50.11	51.20	51.84	53.44
38.70	40.25	42.00	43.15	44.50	45.87	47.19	48.31	49.69	50.82	52.14	53.13	54.45
39.40	41.00	42.70	43.80	45.22	46.71	48.14	49.09	50.52	51.68	53.04	54.06	55.21
40.20	41.75	43.50	44.50	46.06	47.46	48.65	50.05	51.38	52.50	53.90	54.81	56.14
40.75	42.50	44.15	44.90	46.80	48.24	49.60	51.12	52.20	53.28	54.72	55.80	56.88
41.50	43.15	44.85	45.88	47.46	48.84	50.32	51.80	52.91	54.39	55.35	56.36	57.72
42.20	43.85	45.60	46.36	48.03	49.55	51.30	52.44	53.96	55.10	56.24	57.38	58.90
42.80	44.55	46.02	47.19	48.90	50.31	51.63	53.04	54.60	55.77	57.33	58.34	59.67
43.50	44.80	46.40	48.00	49.60	51.20	52.56	54.00	55.44	56.80	58.00	59.20	60.40

enters in the opposite direction (counter flow) at 20 F and leaves the other end of the exchanger at 40 F.

Figure 40 illustrates the log flow.

Find: The logarithm mean temperature difference.

Solution: Table 12 gives LMTD for various temperature differences. Entering the table at 34 and going across to the intersect of 20, the LMTD is 26.5. Or, substituting for Equation 42.

$$\frac{34 - 20}{\log_e 34/20} = \frac{14}{\log_e 1.7} = \frac{14}{0.53} = 26.4$$

12
Balancing Water Flow at the Pump

Waterside balancing begins at the pump, just as air side balancing begins at the fan. Although balancing stations in duct systems are still rare occurrences, their use in water systems are not at all uncommon. When venturi tubes or calibrated orifices are piped into a system the technician's job is simplified considerably. Where such balancing stations are not installed gage cocks and thermometer wells must be provided and installed in such a manner as to enable measuring pressure drop directly across the pump.

Before commencing the water balance, the system must be absolutely clean. Strainers should be cleaned once a week for the first month of operation, once every two weeks for the second month, once a month for the first six months of operation. A clean system is essential to proper balancing of the waterside. After inspecting the strainers and checking the pump rotation, the following must be recorded on a proper form:

1. Suction and discharge pressure at no flow
2. Suction and discharge pressure at full flow
3. Ampere load at no flow
4. Ampere load at full flow
5. With the amps known, the brake horsepower can be calculated from Equation 10;

$$ \text{Bhp} = \frac{\text{running amps} - \text{no load amps} \times 0.5}{\text{full load amps} - \text{no load amps} \times 0.5} \times \text{nameplate hp} $$

6. Pressure drop in feet of head. This is found by multiplying the gage pressure drop in lb/sq in. by 2.31 (See conversion Table 32).

7. With the total head in feet and the brake horsepower known, the gpm may be plotted on the manufacturer's pump curve.

To check the impeller size, close the discharge valve and read the discharge pressure at no load; this should coincide with the curve.

Required Instrumentation

1. Ammeter to check power.
2. Calibrated pressure test gages to check pump pressures.
3. Differential pressure gage to check pressure drops across coils and calibrated orifices.
4. Multipoint surface pyrometer for reading temperatures.

MEASURING FLUID FLOW THROUGH ORIFICE PLATES

Depending upon the geometry of the orifice, the discharge coefficient will vary. The discharge coefficient, designated K, will be supplied by the manufacturer; however, the more common sharp-edged and square-edged orifices will be found to have a coefficient, K = 0.61. Also, d/D will usually be ⅓ or less; "d" being the diameter of the orifice and "D" being the diameter of the pipe containing the orifice. Therefore, the simplified formula is;

$$Q = 19.636 \times d^2 \times 0.61 \sqrt{\Delta h} \qquad (43)$$

where

Q = flow in gpm

d = diameter of orifice

Δh = differential head at the orifice, ft water

If the correction factor K is different than 0.61, the proper K must be substituted. If d/D is greater than 0.3, then multiply gpm (Q) by;

$$0.4 \times 1.014$$

$$0.5 \times 1.033$$

$$0.6 \times 1.070$$

MEASURING FLUID FLOW THROUGH VENTURI TUBES

For any venturi tube in which d = ⅓D. The simplified formula is;

$$Q = 19.17 \times d^2 \sqrt{\Delta h} \qquad (44)$$

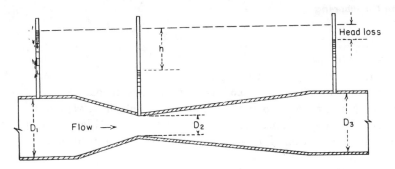

Figure 41. Venturi Meter.

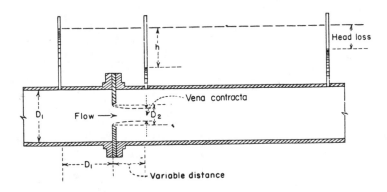

Figure 42. Orifice Meter.

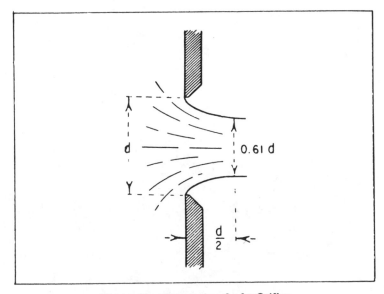

Figure 43. Contraction At An Orifice.

where

    Q = flow in gpm

    d = diameter of venturi throat, in.

  Δh = differential head between upstream end and throat, ft

The manufacturer of the orifice plate or venturi tube usually provides these devices with appropriate pressure taps and instructions. Figs. 41, 42, and 43 illustrate the calculation of d and D for orifice and venturi meters.

13

Balancing Water Flow at the Coil

The usual method of balancing water flow at the coil is to take the pressure drop across the inlet and outlet of the coil and check the drop against the manufacturer's published data; adjusting the water flow to the design. But there are times—on existing jobs—where the published data for a coil are not available. And even on new work, where the data are available, gages and gage cocks are often omitted and there is no ready method of determining the pressure drop. In such cases the water flow across the coil is set by the *wet bulb temperature difference* method.

When balancing water flow we must assume that the air flow has first been balanced, and the system is handling the design cfm. When the cfm is known the Btuh can be found from the equation:

$$\text{Btuh} = \text{cfm} \times 4.5 \times \Delta h \tag{40}$$

where
4.5 is a constant
Δh is the enthalpy difference
The enthalpy difference is found by plotting the entering and leaving wet bulb temperatures on the psychrometric chart. Once the total load in Btuh is known the gpm is calculated from the equation;

$$\text{gpm} = \frac{\text{Btuh}}{500 \times \Delta t} \tag{41}$$

where
500 is a constant (8.33 lb/gal $\times$ 60 min)
Δt is the difference in temperature between the entering and leaving water.

93

Table 13. BTUH Per CFM For Various Wet Bulb Temperature Differentials At Enthalphy Entering Wet Bulb Temperature.

LEAVING WET BULB TEMPERATURE	63	64	65	66	67	68	69	70	71	72	73	74	75	76	77	78	79	80
45	49.1	52.5	55.8	59.3	62.9	66.5	70.2	74.0	77.9	81.8	85.9	90.0	94.3	98.6	103	108	112	117
46	46.8	50.2	53.6	57.0	60.6	64.2	67.9	71.7	75.6	79.5	83.6	87.8	92.0	96.3	101	105	110	115
47	44.5	47.8	51.2	54.7	58.2	61.8	65.6	69.3	73.2	77.2	81.3	85.4	89.7	94.0	98.5	103	108	113
48	42.1	45.5	48.8	52.3	55.8	59.4	63.2	67.0	70.8	74.8	78.9	83.0	87.3	91.6	96.1	101	105	110
49	39.7	43.0	46.4	49.9	53.4	57.0	60.8	64.5	68.4	72.0	76.5	80.6	84.9	89.2	93.7	98.2	103	108
50	37.2	40.5	43.9	47.4	50.9	54.5	58.3	62.1	65.9	69.9	74.1	78.1	82.4	86.7	91.2	95.8	100	105
51	34.7	38.0	41.4	44.9	48.4	52.0	55.8	59.5	63.4	67.4	71.5	75.6	79.9	84.2	88.7	93.2	97.9	103
52	32.1	35.4	38.8	42.3	45.8	49.4	53.1	56.9	60.8	64.8	68.9	73.0	77.3	81.6	86.1	90.6	95.3	100
53	29.5	32.8	36.2	39.6	43.2	46.8	50.5	54.3	58.2	62.1	66.2	70.4	74.7	79.0	83.5	88.0	92.7	97.5
54	26.8	30.1	33.5	36.9	40.5	44.1	47.8	51.6	55.5	59.4	63.5	67.7	72.0	76.3	80.8	85.3	90.0	94.8
55	24.2	27.4	30.8	34.2	37.8	41.4	45.1	48.9	52.8	56.7	60.8	65.0	69.3	73.6	78.1	82.6	87.3	92.1
56	21.3	24.6	28.0	31.5	35.0	38.6	42.3	46.1	50.0	54.0	58.1	62.2	66.5	70.8	75.3	79.8	84.5	89.3
57	18.4	21.7	25.1	28.6	32.1	35.7	39.5	43.2	47.1	51.1	55.2	59.3	63.5	67.9	72.4	77.0	81.6	86.4
58	15.5	18.9	22.1	25.7	29.3	32.9	36.6	40.4	44.2	48.2	52.3	56.4	60.7	65.0	69.5	74.1	78.8	83.6
59	12.6	15.9	19.3	22.7	26.3	29.9	33.6	37.4	41.3	45.2	49.7	53.5	57.7	62.1	66.6	71.1	75.8	80.6
60	9.5	12.8	16.2	19.7	23.2	26.8	30.6	34.3	38.2	42.2	46.3	50.4	54.7	59.0	63.5	68.0	72.7	77.5
61	6.4	9.7	13.1	16.6	20.1	23.7	27.5	31.2	35.1	39.1	43.2	47.3	51.6	55.9	60.4	64.9	69.6	74.4
62	3.2	6.6	9.9	13.4	17.0	20.6	24.3	28.1	32.0	35.9	40.0	44.1	48.4	52.7	57.2	61.8	66.5	71.3
63		3.3	6.7	10.2	13.7	17.3	21.1	24.8	28.7	32.7	36.8	40.9	45.2	49.5	54.0	58.5	63.2	68.0
64			3.4	6.8	10.4	14.0	17.7	21.5	25.4	29.3	33.4	37.6	41.9	46.2	50.7	55.2	59.9	64.7
65				3.5	7.0	10.6	14.4	18.1	22.0	26.0	30.1	34.2	38.5	42.8	47.3	51.8	56.5	61.3
66					3.6	7.2	10.9	14.7	18.5	22.5	26.6	30.7	35.0	39.3	43.8	48.4	53.1	57.9
67						3.6	7.3	11.1	15.0	18.9	23.0	27.2	31.5	35.8	40.3	44.8	49.5	54.3
68							3.7	7.5	11.4	15.3	19.4	23.6	27.9	32.2	36.7	41.2	45.9	50.7
69								3.8	7.7	11.6	15.7	19.8	24.1	28.4	32.9	37.5	42.2	47.0
70									3.9	7.8	11.9	16.1	20.3	24.7	29.2	33.7	38.4	43.2

Table 14. Properties Of Water At Various Temperatures.

DEGREES FAHRENHEIT	DEGREES CELSIUS	DENSITY LBS/FT³	WEIGHT LBS/GAL	VOLUME FT³/LB	SPECIFIC GRAVITY
32	0	62.42	8.34	.01602	1.0
40	4.4	62.42	8.34	.01602	1.0
50	10.0	62.38	8.34	.01603	1.0
60	15.6	62.35	8.33	.01604	1.0
70	21.1	62.27	8.32	.01606	.998
80	26.7	62.19	8.31	.01608	.997
90	32.2	61.11	8.30	.01610	.996
100	37.8	62.00	8.29	.01613	.994
110	43.0	61.84	8.27	.01617	.991
120	49	61.73	8.25	.01620	.990
130	54	61.54	8.23	.01625	.987
140	60	61.40	8.21	.01629	.984
150	66	61.20	8.18	.01634	.980
160	71	61.01	8.16	.01639	.978
170	77	60.00	8.12	.01645	.975
180	82	60.57	8.10	.01650	.970
190	88	60.35	8.07	.01657	.968
200	93	60.13	8.04	.01665	.964
210	99	59.88	8.00	.01670	.960
212	100	59.80	7.99	.01672	.959

Metric Conversion:
Density, lbs/ft³ × 16.0185 = kg/m³
Volume, ft³/lb × 0.0624 = m³/kg
Weight, lbs/gal × 0.11983 = gram/cm³

Example:

A cooling coil has been balanced out on the air side at 10,000 cfm, the entering air reads 69 F wb and the leaving air reads 56 F wb. The entering water temperature reads 45 F db and the leaving water temperature reads 55 F db. Find the gallons per minute circulating through the coil.

Step 1. From the psychrometric chart the entering air at enthalpy saturation is 33.4 and the leaving air is 24.0; therefore, the $\Delta h = 9.4$

Step 2. 10,000 cfm $\times$ 9.4 $\times$ 4.5 = 423,000 Btuh

Step 3. Using the above formula:

$$\text{gpm} = \frac{423,000}{500 \times 10} = 84.6$$

Plotting enthalpy on the psychrometric chart—particularly in the field—is no simple matter. Field conditions, at best, are not conducive to paper work, and the hairline scales of enthalpy saturation are difficult to see clearly even under average lighting conditions let alone a meagerly lit equipment room. Table 13 simplifies the enthalpy calculations and obviates the psychrometric chart for this task.

Applying Table 13 to the above example; the Btuh per cfm for 69 F wb entering air and 56 F wb leaving air is 42.3. Multiplying 42.3 by 10,000 cfm equals 423,000 Btuh.

MEASURING FLUID FLOW THROUGH CONTROL VALVES

The standard measure of valve flow capacity is the valve coefficient, C_v. It is based on the quantity of water—expressed in U.S. gallons per minute (gpm)—that will flow through a wide open valve with a pressure drop of 1 psi. Where the C_v is known the flow in gpm may be measured by the formula;

$$Q = C_v \sqrt{\frac{\Delta P}{G}} \qquad \Delta P = \left(\frac{Q}{C_v}\right)^2 \qquad (45)$$

Q = Flow in U.S. Gallons Per Minute

ΔP = Pressure Drop Through Valve (Inlet Pressure-Outlet Pressure)

G = Specific Gravity (Water = 1.0)

C_v = Valve Flow Coefficient

For hot water application, Table 14 gives the properties of water for various temperatures.

14
Gathering and Recording the Data

To properly record and interpret the test and balancing data, it is necessary to use a complete set of well-designed report forms. Such forms are an essential part of the test report and an important part of the project's history. They are the only method of properly enforcing the specifications and ensuring reliable feedback of empirical data. For the design engineer, these data are essential; they are truly the acid test of his design. If the field data are reliably assembled and presented the engineer will *know* the job was installed in accord with the design. But even more important, with complete and accurate report data, the engineer can verify the design. The engineer can test it in operation, analyze it, accumulate a wealth of practical knowledge, and so gain a new understanding of what to call for in a specification.

A good form will organize the report data in such a manner that it will facilitate the test and balance function itself, and guide the technicians through their tasks in a logical succession of steps—avoiding omissions and errors, and simplifying calculations. As part of the operating and maintenance instructions, which are turned over to the owner with the job, the test and balance report sheets become a valuable part of the permanent record of the system operating conditions and serve as a handy reference for the lifetime of the job.

A typical Duct Traverse Report Sheet is shown in Figure 44.

First the velocity pressure along each point of the traverse is recorded in the V_p column. Each velocity pressure is then converted to velocity, fpm, by referring to Table 10 or Equation 30, and the grand total of velocities is marked on the form (Figure 44, line 5.) It is important to convert each velocity pressure (V_p) to an equivalent velocity (V) and then take the average of the sum of the V's rather than taking the average of the sum of the V_p's and converting into an average V. The sum divided by the number of reading points equals the average fpm.

DUCT TRAVERSE READINGS

PROJECT _F. W. D Muller_   SHEET _/_ OF _/_

SYSTEM _West Building_ _____ FLOOR # _4_   ZONE # _T_

JOB # _614_   DATE _Oct 3 1979_

REMARKS _Bill Indiana & Jack Craft / Team 2_

NO.	VELOCITY PRESSURE	VELOCITY	DUCT POSITION	NO	VELOCITY PRESSURE	VELOCITY	DUCT POSITION	NO.	VELOCITY PRESSURE	VELOCITY	DUCT POSITION
1A	.056	946		1B				1C			
2A	.068	1045		2B				2C			
3A	.028	670		3B				3C			
4A	.078	1119		4B				4C			
5A	.081	1140		5B				5C			
6A	.084	1161		6B				6C			
7A	.086	1175		7B				7C			
8A	.088	1188		8B				8C			
9A	.087	1181		9B				9C			
10A	.083	1154		10B				10C			
11A	.080	1133		11B				11C			
12A	.088	1188		12B				12C			
13A	.072	1075		13B				13C			
14A	.052	913		14B				14C			
15A	.052	913		15B				15C			
16A	.051	904		16B				16C			
TOTAL _16,905_ ÷ _16_				TOTAL _____ ÷ ___				TOTAL _____ ÷ ___			
= AVERAGE VELOCITY _1056_				= AVERAGE VELOCITY _____				= AVERAGE VELOCITY _____			

1. TEST NUMBER _____ _1_ _____
2. ALTITUDE ____ _5000 FT._ ____ = CORRECTION FACTOR _1.10_
3. AIR TEMPERATURE _450°F_ _____ = CORRECTION FACTOR _1.31_
4. COMBINED FACTOR (LINE 2 X LINE 3) _____ = _____ _1.441_
5. GRAND TOTAL VELOCITIES _16,905_ = _1056_ V_m, AVERAGE
   NUMBER OF READING POINTS _16_
6. V_m _1056_ (LINE 5) X FACTOR _1.441_ (LINE 4) = _1522_ V , FPM
7. DUCT AREA _6_ FT^2
8. V _1522_ (LINE 6) X A _6_ (LINE 7) = _9132_ CFM
9. ACTUAL CFM _9132_ (LINE 8)   DESIGN CFM _____ _____
10. STATIC PRESSURE (S_p) AT CENTER OF TEST ____ _.086_ _____

   V_m = MEASURED VELOCITY, OR INDICATED VELOCITY
   V = ACTUAL VELOCITY AFTER DENSITY VARIATION CORRECTION
   A = DUCT AREA, FT^2
   RECTANGULAR DUCT AREA, FT^2 = WIDTH X HEIGHT, IN./ 144
   ROUND DUCT AREA, FT^2 = 3.142 X RADIUS SQUARED (πR^2)
   FINAL CORRECTION FACTOR FOR NON-STANDARD AIR = ALTITUDE
      FACTOR X TEMPERATURE FACTOR.

Figure 44. Duct Traverse Report Sheet.

AIR HANDLER TEST REPORT

PROJECT _____

SYSTEM _____ FLOOR # _____

JOB # _____

REMARKS _____

SHEET _____ OF _____

ZONE # _____

DATE _____

I T E M	SPECIFIED	FIELD TEST 1	FIELD TEST 2	FIELD TEST 3
Outside air cfm				
Total air cfm				
% outside air				
Discharge duct sq. ft.				
Discharge duct fpm				
Return duct sq. ft.				
Return duct fpm				
Return air cfm				
Manufacturer				
Fan size				
Arrangement				
Fan blade				
Fan sheave				
Motor sheave				
No. rows coil				
Filters				
Rpm				
Hp				
Bhp				
Volts				
Phase				
Cycle				
Full-load amps				
No-load amps				
Heaters: rated amps				
Suction sp				
Discharge sp				
Total sp				
Ent. DB temperature				
Ent. WB temperature				
Lvg. DB temperature				
Lvg. WB temperature				
Outside air temperature				
Gpm circulating H_2O				
Pressure drop				
Ent. water temperature				
Lvg. water temperature				

Remarks _____

Figure 45. Air Handler Test Report Sheet.

The Air Handler Test Report shown in Figure 45 is another example of a well designed form for recording and organizing data necessary for a complete test and balance report.   Available, too, are forms for terminal outlets, chillers, hydronic systems, energy recovery equipment, and so forth.   These are offered through such associations and groups as the *Associated Air Balance Council, SMACNA*, and *American Conference of Governmental Industrial Hygienists*.   Manufacturers of balancing stations, venturi flow meters and balancing instrumentation may also provide a variety of test report forms and air balancing work sheets.   *The National Environmental Balancing Bureau*[1] offers 27 separate test and balance report forms (8 × 10 in., in pads of 50).

[1]National Environmental Balancing Bureau, 8224 Old Courthouse Road, Vienna, Virginia 22180.

15
Variation in Air Density

A common error in testing adjusting and balancing of air systems is the neglect of correcting data for non-standard air conditions, i.e. air other than 70 F and 29.92 barometric pressure. Process exhaust and drying systems, commercial kitchen exhaust, and warm air systems for thermal comfort demand exacting attention to problems of air density variation.

The equation that relates the heat carrying capacity of air to temperature has been stated earlier; Btuh = WC Δ t (34). How W relates to cfm may best be expressed by:

$$W = cfm \times 60 \times \text{air density} \qquad (46)$$

where
$\qquad$ W = air-flow rate in pounds of air per hour.
$\quad$ air density = the weight of one ft^3 of air.
Clearly, it is the weight of the air that determines the heat transfer rate; the constant, 1.08 can apply only to standard air with a density of 0.075 pounds per cubic foot (0.075 $\times$ specific heat, 0.24 $\times$ 60 min= 1.08). Since the density of air is not constant but varies with changes in barometric pressure and temperature all calculations must be corrected to reflect the actual air density.

Unfortunately, long ago the industry inaugurated a system of stating air-flow rates in cubic feet per minute rather than in pounds per hour. Obviously the cfm will vary as the air enters a coil at say, 70 F and leaves at 200 F, a complication we would not be faced with if the air-flow were stated in lb per hour because the pounds per hour would remain constant at all points in the system.

Table 15. Temperature–Density Corrections For Dry Air.
Atmospheric pressure 29.92 in. Hg (101 kPa)

1	2	3	4	5	6
DEGREES FARENHEIT	DEGREES CELSIUS	DENSITY LBS/FT^3	VOLUME FT^3/LB	DENSITY RATIO	CORRECTION FACTOR*
32	0	.0807	12.38	1.08	0.96
40	4.4	.0794	12.59	1.06	0.97
70	21.1	.0749	13.34	1.00	1.00
100	37.8	.0709	14.10	.95	1.03
150	66.0	.0651	15.36	.87	1.07
200	93	.0602	16.62	.80	1.12
250	121	.0559	17.88	.75	1.15
300	149	.0523	19.13	.70	1.20
350	177	.0490	20.39	.65	1.24
400	204	.0462	21.65	.62	1.27
450	232	.0430	22.98	.58	1.31
500	260	.0414	24.17	.55	1.35
550	288	.0392	25.48	.52	1.39
600	316	.0375	26.69	.50	1.41
650	343	.0358	27.95	.48	1.44
700	371	.0342	29.21	.46	1.47
750	399	.0328	30.47	.44	1.51
800	427	.0315	31.73	.42	1.54
850	460	.0303	32.99	.40	1.58
900	482	.0292	34.24	.39	1.60
950	516	.0282	35.51	.37	1.64
1000	538	.0272	38.76	.34	1.71

*To correct for nonstandard air, multiply the correction factor from column 6 by the Pitot tube measured velocity.

$$V = V_m \times \text{correction factor}$$

To correct for non-standard conditions not shown in this table use Equation 49:

$$V = V_m \sqrt{\frac{0.075}{d}}$$

where
   V = actual velocity (corrected)
   V_m = measured velocity
   d = actual air density

Table 16. Altitude–Density Corrections For Dry Air.
Air at 70 F (21.1 C)

1	2	3	4	5	6
ALTITUDE, FEET	INCHES OF MERCURY	DENSITY LBS/FT^3	VOLUME FT^3/LB	DENSITY RATIO	CORRECTION FACTOR*
-1000	31.02	.0775	12.90	1.03	0.98
Sea level	29.92	.0749	13.69	1.00	1.00
500	29.39	.0735	13.60	.98	1.01
1000	28.86	.0721	13.87	.96	1.02
1500	28.33	.0708	14.12	.95	1.03
2000	27.82	.0695	14.39	.93	1.04
3000	26.81	.0670	14.92	.89	1.06
4000	25.84	.0646	15.48	.86	1.08
5000	24.89	.0622	16.08	.83	1.10
6000	23.98	.0600	16.67	.80	1.12
7000	23.09	.0577	17.33	.77	1.14
8000	22.22	.0550	18.18	.73	1.17
9000	21.38	.0530	18.87	.71	1.19
10,000	20.58	.0510	19.61	.68	1.21

*To correct for non-standard air, multiply the correction factor from column 6 by the Pitot tube   measured velocity.

$$V = V_m \times \text{correction factor}$$

To correct for non-standard conditions not shown in this table use the formula:

$$V = V_m \sqrt{\frac{0.075}{d}}$$

where
   V = actual velocity (corrected)
   V_m = measured velocity
   d = actual air density

Example:

Given: A heat carrying computation requires an air-flow of 2,000 lb/hr.
Find: The air quantity in cfm of standard air.

$$\textit{Solution:} \quad \frac{\text{lb/hr}}{0.075 \times 60 \text{ min}} = \text{cfm} = \frac{2000}{0.075 \times 60 \text{ min}} = \frac{2000}{4.5} = 444 \text{ cfm}$$

Table 17. Barometric Pressure–Density Correction For Dry Air.

DEGREES FAHRENHEIT	BAROMETRIC PRESSURE, INCHES OF MERCURY			
	29.92*	29.50	29.00	28.50
	CORRESPONDING AIR DENSITIES			
0 F	0.086	0.085	0.084	0.082
10 F	0.085	0.083	0.082	0.080
20 F	0.083	0.082	0.080	0.079
30 F	0.081	0.080	0.079	0.077
40 F	0.079	0.078	0.077	0.076
50 F	0.078	0.077	0.075	0.074
60 F	0.076	0.075	0.074	0.073
70 F	0.075	0.074	0.073	0.071
80 F	0.074	0.073	0.071	0.070
90 F	0.072	0.071	0.070	0.069
100 F	0.071	0.070	0.069	0.068
110 F	0.070	0.069	0.068	0.065
120 F	0.069	0.068	0.066	0.065
130 F	0.067	0.066	0.065	0.064
140 F	0.066	0.065	0.064	0.063
150 F	0.065	0.064	0.063	0.062
160 F	0.064	0.063	0.062	0.061
170 F	0.063	0.062	0.061	0.060
180 F	0.062	0.061	0.060	0.059
190 F	0.061	0.060	0.059	0.058
200 F	0.060	0.059	0.058	0.057
210 F	0.059	0.058	0.057	0.056
220 F	0.058	0.058	0.057	0.056
230 F	0.058	0.057	0.056	0.055
240 F	0.057	0.056	0.055	0.054

*29.92 is standard barometric pressure

$$\text{Density} = 1.325 \times \frac{\text{barometric pressure, in. Hg}}{\text{air temperature, F plus 460 F}}$$

Correction factors may be extracted from corresponding air densities found in either Table 15 or 16, or from Equation (48).

Find: The air quantity in cfm if the air is 200 F and sea level.
Solution: The density of 200 F air is 0.062; therefore,

$$\frac{2000}{0.062 \times 60 \text{ min}} = \frac{2000}{3.72} = 538 \text{ cfm}$$

CORRECTING FOR NON-STANDARD AIR

At elevated temperatures or altitudes the volume of air will increase as the reciprocal of the density ratio—a fact often overlooked by the designer and the air balance technician.   Tables 15, 16, and 17 give air density and correction factors for corresponding temperatures, altitudes and barometric pressures. The Speed-O-Graphs, Figures 46 and 47 are convenient tools for finding the air density ratio quickly.   To determine the dry air density by formula:

$$d = 1.325 \frac{P_b}{T} \qquad (47)$$

where
  d = air density, lb/ft.^{3}
  P_b = barometric pressure, in Hg
  T = absolute temperature (F plus 460)
  The correction factors in Tables 15 and 16 are precalculated values taken from the ratio of the density of air at non-standard conditions to the density of the air at standard conditions.

$$\text{Correction factor} = \sqrt{\frac{0.075}{d}} \qquad (48)$$

Where the correction factor has been determined, it is multiplied directly by the measured velocity to solve for the true, or actual, velocity.

$$V = V_m \sqrt{\frac{0.075}{d}} \qquad (49)$$

where
  V = actual velocity (corrected)
  V_m = measured velocity
  d = actual air density of the measured air
Approximate correction for non-standard air may be made quickly by allowing a 2% increase in velocity pressure for each 10 degrees above 70 F and 4% for each 1000 feet altitude above sea level.

Example:

  Given: At sea level application, 200 F air is exhausted from a process. The velocity pressures across the traverse average 0.09 V_p.

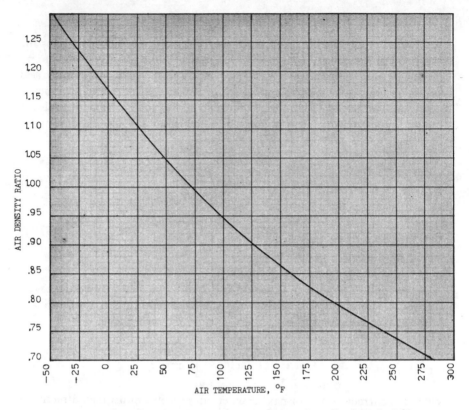

Figure 46. Speed-O-Graph: Temperature Corrections For Standard Air.

To find the correction factor, take the square root of the reciprocal of the air density ratio. Correction factor = $\sqrt{1/d_r}$. Multiply the correction factor by the measured velocity to find the actual V. For example, air at 175 F and sea level, find the density ratio = 0.83. Then, $\sqrt{1/0.83}$ = 1.10. If the measured air is 3200 fpm, then 3200 × 1.10 = 3520 fpm, actual velocity.

$$\text{Air density ratio} = \frac{\text{density of the measured air}}{0.075} = d_r$$

Find: The actual velocity in fpm
Solution: 1. 200 F − 70 F = 130 F difference
2. 130 F / 10 = 13 F
3. 13 × 0.02 = 0.26
4. 0.09 V_p × 1.26 = 0.1134 V_p corrected

$$V = 4005 \sqrt{0.1134} = 1349 \text{ fpm}$$

Example:

Given: At an altitude of 2000 feet above sea level a system is delivering 70 F air. The velocity pressures read 0.20 V_p across the traverse.

Find: The actual air velocity in fpm.

Solution: 1. 2000 ft = 8%
                2. 0.20 V_p × 1.08 = 0.216 V_p
                3. V = 4005 $\sqrt{0.216}$ = 1861 fpm

When testing an air system at both elevated temperature and altitude, the correction factors must be multiplied together.

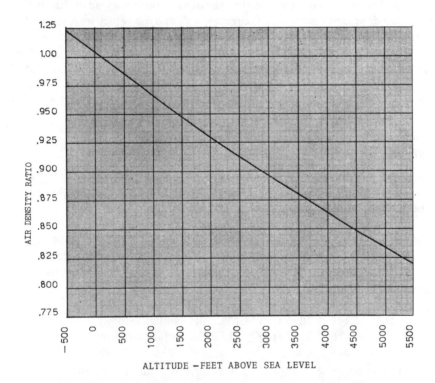

Figure 47. Speed-O-Graph: Altitude Corrections For Standard Air.

To find the correction factor, take the square root of the reciprocal of the air density ratio. Correction factor = $\sqrt{1/d_r}$. Multiply the correction factor by the measured velocity to find the actual V. For example, air at 5500 ft above sea level, find air density ratio = 0.820. Then, $\sqrt{1/0.820}$ = 1.22. If the measured velocity is 1500 fpm then 1500 × 1.22 = 1820 fpm, actual velocity.

Example:

An exhaust system shows an indicated velocity of 3000 fpm. The system is being tested at an altitude of 4000 ft and the air temperature measures 250 F.

Find: The actual corrected air velocity.

Solution: A. From Table 15, column 6, the temperature correction factor for 250 F is 1.15.

             B. From Table 16, column 6, the altitude correction factor for 4000 ft above sea level is 1.08.

             C. Multiplying both factors, gives $1.15 \times 1.08 = 1.242$

             D. 3000 fpm $V_m \times 1.242 = 3726$ fpm, actual velocity.

Among the several ways of correcting for non-standard air are charts, slide rules, and formulas. The reciprocal of the density ratio may be multiplied by the measured velocity pressure to correct the V_p and the velocities totaled. Or, the square root of the density ratio may be multiplied by the average velocity *before* correcting from the V_p to find the corrected velocity. That is to say the correction may be made for V_p and converted to V, or the V_p may be converted to V and then corrected. The fastest and most reliable method is the latter. The following example gives the step-by-step procedure for reading, correcting and recording the Pitot traverse.

Example:

A duct system at the Pitot tube station is 36 × 24 in. and the measurements are made at an altitude of 5000 ft with duct air temperatures of 450 F. Assume a 16 point Pitot tube reading. (See table, page 109)

Substituting for $Q = AV$ and 6 ft^2 duct, $6 \times 1056 = 6336$ cfm standard air. But correcting for non-standard conditions from Table 15, the factor is 1.31 and from Table 16, the factor is 1.10; therefore $1.31 \times 1.10 = 1.441$ for both temperature and altitude.

$$1.441 \times 1056 \text{ fpm} = 1522 \text{ fpm, actual velocity}$$
$$6 \times 1522 \text{ fpm} = 9132 \text{ cfm, actual air-flow}$$

Table 17A shows the effects of temperature on moving air in relation to changes in air density.

SUMMARY

To correct for non-standard air conditions, proceed with the Pitot tube traverse as if it were standard air.

PITOT TUBE STATION	VELOCITY PRESSURE IN. WG	VELOCITY, FPM STANDARD AIR
1A	.056	946
2A	.068	1045
3A	.028	670
4A	.078	1119
5A	.081	1140
6A	.084	1161
7A	.086	1175
8A	.088	1188
9A	.088	1188
10A	.087	1181
11A	.083	1154
12A	.080	1133
13A	.072	1075
14A	.052	913
15A	.052	913
16A	.051	904

GRAND TOTAL VELOCITIES = 16,905

AVERAGE VELOCITY $= \dfrac{16,905}{16} = 1056$ fpm

1. Record the velocity pressures, V_p for each point.
2. Convert each V_p to velocity in fpm.
3. Total all the velocities.
4. Divide the total of velocities by the number of Pitot tube readings to find the average velocity.
5. *Multiply the average velocity by the correction factors (found in Tables 15, 16, 17 or Figures 46 and 47) to find the actual corrected velocity for the variance in air density.*
6. Multiply the corrected velocity by the duct area in square feet to find the actual air-flow in cfm.

To find the barometric pressure where the elevation is known use the approximate correction of 0.1 inch pressure reduction per 100 ft elevation.

Table 17A. Effect Of Temperature On Moving Air.

(1) TEMPERATURE DEGREES F	(2) VOLUME AT SAME WEIGHT	(3) VELOCITY TO PRODUCE SAME PRESSURE	(4) PRESSURE TO MOVE SAME WEIGHT, ORIFICE	(5) FAN SPEED SAME WEIGHT, ORIFICE	(6) POWER, SAME WEIGHT, ORIFICE	(7) POWER, SAME VOLUME, RESISTANCE	(8) POWER, SAME WEIGHT, RESISTANCE	(9) POWER, SAME VOLUME, VELOCITY
25	.916	.957	.916	.916	.839	1.00	.916	1.09
30	.926	.962	.926	.926	.857	1.00	.926	1.08
35	.935	.967	.935	.935	.874	1.00	.935	1.07
40	.945	.972	.945	.945	.893	1.00	.945	1.06
45	.954	.976	.954	.954	.910	1.00	.954	1.05
50	.962	.981	.962	.962	.925	1.00	.962	1.04
55	.972	.986	.972	.972	.945	1.00	.972	1.03
60	.983	.991	.983	.983	.966	1.00	.983	1.02
65	.992	.996	.992	.992	.984	1.00	.992	1.01
70	1.00	1.000	1.00	1.00	1.00	1.00	1.00	1.00
75	1.01	1.005	1.01	1.01	1.02	1.00	1.01	.99
80	1.02	1.010	1.02	1.02	1.04	1.00	1.02	.98
85	1.03	1.015	1.03	1.03	1.06	1.00	1.03	.97
90	1.04	1.019	1.04	1.04	1.08	1.00	1.04	.96
95	1.05	1.025	1.05	1.05	1.10	1.00	1.05	.95
100	1.06	1.029	1.06	1.06	1.12	1.00	1.06	.94
110	1.08	1.039	1.08	1.08	1.17	1.00	1.08	.93
120	1.10	1.049	1.10	1.10	1.21	1.00	1.10	.91
125	1.11	1.053	1.11	1.11	1.23	1.00	1.11	.90
150	1.15	1.072	1.15	1.15	1.32	1.00	1.15	.87
175	1.20	1.095	1.20	1.20	1.44	1.00	1.20	.83

(1)	(2)	(3)	(4)	(5)	(6)	(7)	(8)
200	1.25	1.118	1.25	1.56	1.00	1.25	.80
225	1.29	1.135	1.29	1.66	1.00	1.29	.78
250	1.34	1.157	1.34	1.80	1.00	1.34	.75
275	1.39	1.179	1.39	1.93	1.00	1.39	.72
300	1.43	1.196	1.43	2.04	1.00	1.43	.70
325	1.48	1.216	1.48	2.19	1.00	1.48	.68
350	1.53	1.237	1.53	2.34	1.00	1.53	.65
375	1.58	1.257	1.58	2.50	1.00	1.58	.63
400	1.62	1.273	1.62	2.62	1.00	1.62	.62
425	1.67	1.292	1.67	2.79	1.00	1.67	.60
450	1.72	1.311	1.72	2.96	1.00	1.72	.58
475	1.77	1.330	1.77	3.13	1.00	1.77	.57
500	1.81	1.345	1.81	3.28	1.00	1.81	.55

The effect of temperature is shown on the volume, weight, pressure, and power required to move dry air at atmospheric pressure of 29.92 inches of mercury absolute. Column (1) gives the temperature in degrees F. Column (2) shows the relative volume of the air at the same weight, and column (3) gives the relative velocities that would be required to develop the same pressure. Column (4) gives the relative pressure required to move the air at the same weight through the same orifice and column (5) gives the relative rpm to move the air at the same weight through the same orifice. Column (6) shows the relative horsepower required to move the air at the same orifice and column (7) gives the relative horsepower required to move the air at the same weight through the same orifice. Column (8) shows the relative horsepower needed to move the same volume of air against the same resistance and column (9) gives the relative power needed to move an equal volume of air at the same velocity. The baseline for these data is 70F. The ratio for 70F in all eight cases is 1.00.

For example, if the elevation is known to be 1780 ft above sea level then, 1780 ft / 100 ft × 0.1 in. = 1.78 in.; therefore, the barometric pressure will be 29.92 − 1.78 = 28.14 inches of mercury. This may be used to substitute in Equation (47) to find the actual air density for ranges not normally given in the tables.

16
Testing Exhaust Hoods

Thermal comfort systems have their own built-in, closed-loop, feedback circuits, so to speak, in the form of complaints from occupants of the treated space. The criterion for any comfort system is, after all, an expression of satisfaction from the space occupants regardless of the theoretical opinions of engineers and test and balance technicians. That is why an intelligent and skilled test and balance technician can balance a space by taking a room traverse with a sling psychrometer rather than cfm readings at the outlets. Because of the feedback of human response to cooling and heating systems, engineers, test and balance personnel, servicemen, building managers, and others are constantly adjusting and correcting system faults which would otherwise never be recognized.

Although thermal comfort is often a consideration in exhaust hood design the main function of such a system is the removal of noxious fumes, air pollutants, smoke, odors, grease, etc. The extent to which such a system has been correctly designed and properly operated can only be measured by the professional test and balance team; there is no natural feedback system of complaints. Consequently, the most critical problems regarding exhaust hood design are left undetected and unattended. The gravity of this condition is most obvious in the restaurant industry where the fire loss statistics of restaurants and road-side eating establishments has risen to serious proportions. Thousands of restaurants have gone to cinders and ashes unnecessarily. It has been observed that only a rare few of the total cooling and heating jobs installed in this country are subjected to serious testing, balancing, and adjustment. As for restaurant exhaust systems, practically none are tested, adjusted, or balanced.

Kitchen exhaust hoods and ventilating systems are seldom designed by engineers. In those few cases where engineers are employed to design these systems, the greatest design effort seems to go toward fire protection, rather than fire prevention, and the average air conditioning engineer has neither the experience nor the feedback data to improve his knowledge of the thermo-dynamics of kitchen hood exhaust and hood design.

Fire codes are of little help. In fact, because the fire codes set minimum standards, which quickly tend to become maximum standards in practice, the fire codes are often harmful. Experience has demonstrated that fire codes have not slowed the rate of restaurant kitchen fires. Code requirements universally accept the method of relating the exhaust air volume (cfm) to the face area of the hood. Engineers who have adopted this rule-of-thumb method for kitchen exhaust seem to have forgotten the equation that relates the heat carrying capacity of air to temperature: $Btu = WC \, \Delta t$, and the relationship of W to air density.

It has been observed that operating duct temperatures in restaurant kitchen exhaust systems may approach 300 F. Where ducts are lined with grease—and that should always be assumed in restaurant application—the control of duct temperatures is of critical importance to fire prevention. With the introduction in recent years, of heavy batteries of high heat producing appliances in fastfood operations, particular attention must be given to problems of *heat removal* and *keeping the ducts cool*. Air-flow rates based only on area relationships and hood capture velocities without regard to appliance and duct temperatures are imprecise at best.

APPROACHING THE KITCHEN EXHAUST HOOD

Since the first edition in 1951 of *Industrial Ventilation*, the American Conference of Governmental Industrial Hygienists has updated its' important manual 14 times. This manual should be on the desk of every design engineer and test and balance technician who may be involved with ventilation and exhaust projects; but it does not discuss kitchen exhaust in depth. In fact, there is no single source of in-depth reference for kitchen exhaust design and system testing. The design of exhaust systems whether for industrial processes or commercial and institutional restaurant kitchens, is not the concern of our particular work, but owing to the lack of information and generally poor design of such systems, the test and balance technician needs to be familiar with these kinds of problems. The following rules will help the test and balance technician approach kitchen exhaust projects dialectically and confidently:

1. Develop a good report form. Figure 48 shows the form recommended by the American Conference of Governmental Hygienists. It is not specifically addressed to kitchen exhaust hoods, but it will work. It is

Plant_____ Dept._____ Date_____
Operation exhausted_____ By_____

Line sketch showing points of measurement

Date system installed_____

Hood and transport velocity

Point	Duct		VP in. H$_2$O	SP in. H$_2$O	FPM (Fig.6-16)	CFM Q = VA	Remarks
	D	Area (Fig.6-18)					

Pitot traverse
Pitot readings – See Fig. 6-16 & Tables 9-1, 9-2, 9-3

Points	VP	Vel.	VP	Vel.	VP	Vel.
1						
2						
3						
4						
5						
6						
7						
8						
9						
10						
Total Vel.						
Average Vel.						
CFM						

Fan
Type_____
Size_____

Point	Dia.	SP	VP	TP	CFM
Inlet					
Outlet					

Fan SP_____(See Section 6)

Motor
Name_____ Size_____
HP_____ E_____ I_____ W_____

Collector
Type & size_____

Point	Dia.	SP	Δ SP
Inlet			
Outlet			

Notes_____

Figure 48. Exhaust Hood Report Sheet.

significant to note that the National Environmental Balancing Bureau does not include an exhaust hood report sheet among their 27 standard test and balance report forms.

2. Expect the worst. Kitchen exhaust hoods are rarely well-designed. The Uniform Mechanical Code, Standard Mechanical Code, and NFPA 96 simply require that the minimum air velocity through any duct shall be 1500 fpm. At best, they may require an air quantity of 150 cfm per ft^2

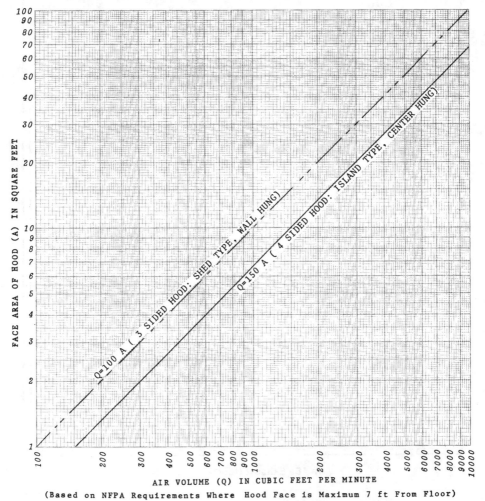

(Based on NFPA Requirements Where Hood Face is Maximum 7 ft From Floor)

Figure 49. Speed-O-Graph: Air Volume Required For Canopy-type Kitchen Exhaust Hoods.

area of hood face (see Figure 49)—a criterion which most designers accept but which has no relationship to heat quantities or intensities.

3. Expect the worst. Kitchen exhaust systems are usually sized for standard air. Air temperature densities are ignored resulting in fan pressures as much as 40% below actual requirements.

4. Expect the worst. Underventilation, overventilation, and inadequate make-up air are prevalent in the great majority of systems operating.

5. Expect the worst. Poorly fabricated duct fittings; restricted make-up air; worn, slipping belts; and overloaded impingement-type grease filters are all too common occurrences which even the most seasoned designer does not allow for, but which can bring a system to its autoignition point sending the entire establishment up in flames.

6. Expect the worst. Most designers follow manufacturers recommendations by allowing 0.20 in. static pressure for filter grease build-up. It is not unusual for the actual build-up to go to 0.40 or 0.50, or even 0.80 in. of static pressure.

7. Expect the worst. Owing to the inaccessibility and difficulty of taking Pitot tube traverses under cramped, hot conditions, kitchen exhaust systems are the most logical places for orifice meters and built-in

Industrial Ventilation, 13th ed. lists the following standards for kitchen range hoods, etc.

TYPE HOOD	AIR QUANTITY CFM	ENTRY LOSS, INCHES WG	DUCT VELOCITY FPM
1. Hood against wall, 3 sides	80/ft² of hood area	0.25 (Filters), +0.50 duct VP	1000 to 4000
2. Island type 4 sides	125/ft² of hood area	0.25 (Filters), +0.50 duct VP	1000 to 4000
3. Low side, wall hood	200/lineal ft of cooking surface	Filters + 0.25 duct VP	1000 to 4000
4. Canopy Dishwasher	250/ft² of face area	0.25 duct VP	1000 to 3000
5. Slot Dishwasher	150/ft² of door area	1.00 slot VP +0.25 duct VP	1000 to 3000
6. Vestibule Dishwasher	150/ft entrance and exit area	0.50 duct VP	1000 to 3000
7. Charcoal Broiler	100 × length × width of front	0.25 (Filters) +0.25 duct VP	1000 to 3000
8. Barbecue Pits	100 × width × height of front	0.25 (Filters) +0.25 duct VP	1000 to 3000

These are minimum standards to be observed and reported.

measuring stations, but the author is aware of no commercial kitchen system where an orifice meter can be found. Pitot tube readings will have to be made.

FOUR BASIC STEPS FOR BALANCING EXHAUST HOODS

1. Observe, adjust, and record the capture velocity patterns around the hood. Use a small smoke generator for this task.
2. Measure, balance, and record the air-flow across the hood face or opening. Use a heated thermocuple anemometer for this task.
3. Measure, balance, and record the exhaust air-flow at the terminal point (roof or other location). Note the proximity of outdoor air intakes, louvers or windows to the point of contamination release, and report same. The exhaust air-flow should be measured by Pitot tube traverse in the exhaust duct at a point close to the discharge fan.
4. Measure, balance, and record the supply air to the space containing the hood. Note all air systems which interface the exhaust system, recording air quantities, velocities, and locations of air streams in relation to the exhaust system.

As a final step, the capture velocity patterns around the hood should be once again observed and noted. Complete fan data for all exhaust and make-up air fans must be collected and recorded.

Observations should be made for both hot and cold fan operation and corrections for non-standard air *must* be entered.

Where exhaust hoods are used in spaces cooled by variable volume air supply systems, it will be very difficult to balance the space. Such conditions must be brought to the attention of the design engineer, or in the case of existing buildings, the plant manager or plant engineer.

17
Troubleshooting Fans

NOISE

The most common fan problem is noise. The experienced technician can usually trace the source of mechanical or vibrational noise quickly. A stethoscope should be the only instrument needed. If the problem cannot be isolated, the possibility of surge, resonance, or unbalance must then be checked in that sequence.

Surge is a common condition that occurs when a fan is operating somewhere in the unstable area to the left of its fan curve. It is a static pressure problem and must be relieved accordingly by reducing the system static. Duct blockage, jammed fire dampers, plugged filters, or improper fan selection can be the cause. A discharge-to-suction bypass may be the solution. Reducing the fan speed will not eliminate surge.

Resonance is usually caused by transmitted vibration to the apparatus panel. By changing the fan speed in either direction by more than 10%, the resonance will vanish. The correction or alteration can then be made by damping, absorbing, or stiffening the faulty panel.

If none of the above are found to be the case the problem may be *unbalance*. This is a more serious fan problem—the fan must be balanced. The only acceptable instrument for this task is an electronic vibration analyzer.

Another frequent cause of unstable fan operation is fully depressed scroll volume dampers in a shipping position. Check the scroll volume dampers before proceeding with the test.

PERFORMANCE REDUCTION

There are two conditions often eluding both engineer and test and balance technician which may cause poor fan performance: fan inlet restriction and fan inlet spin.

119

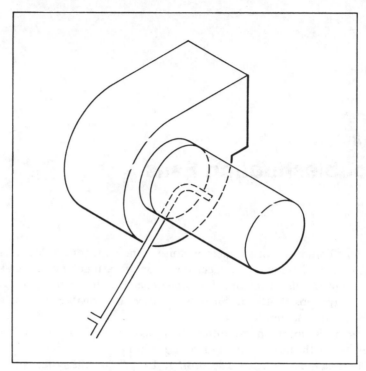

Figure 50. Probing For Spin.

As a consequence of job conditions or poor system engineering, a fan may be located too close to a wall causing a serious reduction in performance. Where a fan inlet is closer than one wheel diameter from an adjoining wall, it is suggested that the fan be inspected for performance on its curve. Fan performance may also be seriously impeded by a drive guard restriction. Belt and drive guards should be kept as far as possible from the inlet opening, and, wherever possible, fabricated from expanded metal mesh rather than solid sheet steel. Be alert for fan *inlet restrictions*.

Inlet spin, resulting from a poor inlet connection, is probably the most frequent cause of poor fan performance. When air is introduced into a fan plenum it should be directed at the center-line of the fan inlet. If the return air velocity spin is imparted in the direction of wheel rotation, the fan volume, pressure, and horsepower are lowered. If the air spin is opposite to the wheel rotation, the volume, static pressure, and horsepower will be greater than expected. Either situation of spin will cause a reduction in fan efficiency.

Frequently, the designer will neglect the fan inlet connection.

It is hardly uncommon to find that the designer has failed to consider all sides of the fan inlet connection. If straight air flow is imparted at the inlet connection the inlet tangential velocity will be zero; any other angle of velocity at the inlet will affect the fan performance.

CHECKING FOR SPIN

By inserting a Pitot tube through the flexible connection at the fan inlet it is easy to demonstrate the condition of spin. With the Pitot tube connected to a draft gage or inclined manometer to read velocity pressure,[1] probe the connection carefully, holding the Pitot tube parallel to the fan shaft, Figure 50. Eccentric flow will be indicated by higher readings on the top, bottom, or side of the connection. The angle of air flow will be indicated by slowly oscillating the Pitot tube back and forth. By carefully observing the pressure readings as the Pitot tube is slowly twisted back and forth, it is possible to determine the angle of spin. To figure the loss in capacity equivalent to the angle of spin, the technician may consult a reliable inlet damper characteristic curve.

The following Master Troubleshooting Chart has been extracted from the Air Movement and Control Association (AMCA), *Fan Application Manual*, with permission.

A. NOISE

Source	Probable Cause
1. IMPELLER HITTING INLET RING	a. Impeller not centered in inlet ring. b. Inlet ring damaged. c. Crooked or damaged impeller. d. Shaft loose in bearing. e. Impeller loose on shaft. f. Bearing loose in bearing support.
2. IMPELLER HITTING CUTOFF	a. Cutoff not secure in housing. b. Cutoff damaged. c. Cutoff improperly positioned.
3. DRIVE	a. Sheave not tight on shaft (motor and/or fan). b. Belts hitting belt tube. c. Belts too loose. Adjust for belt stretching after 48 hours operating.

[1] The static opening connects to the low side of the manometer and the total pressure opening connects to the high side.

A. NOISE

Source	Probable Cause
	d. Belts too tight.
	e. Belts wrong cross section.
	f. Belts not "matched" in length on multi-belt drive.
	g. Variable pitch sheaves not adjusted so each groove has same pitch diameter (multi-belt drives).
	h. Misaligned sheaves.
	i. Belts worn.
	j. Motor, motor base or fan not securely anchored.
	k. Belts oily or dirty.
	l. Improper drive selection.
4. COUPLING	a. Coupling unbalanced, misaligned, loose, or may need lubricant.
5. BEARING	a. Defective bearing.
	b. Needs lubrication.
	c. Loose on bearing support.
	d. Loose on shaft.
	e. Seals misaligned.
	f. Foreign material inside bearing.
	g. Worn bearing.
	h. Fretting corrosion between inner face and shaft.
6. SHAFT SEAL SQUEAL	a. Needs lubrication.
	b. Misaligned.
7. IMPELLER	a. Loose on shaft.
	b. Defective impeller. *Do not run fan. Contact manufacturer.*
	c. Unbalance.
	d. Coating loose.
	e. Worn as result of abrasive or corrosive material moving through flow passages.
8. HOUSING	a. Foreign material in housing.
	b. Cutoff or other part loose (rattling during operation).
9. ELECTRICAL	a. Lead-in cable not secure.
	b. AC hum in motor or relay.

A. NOISE

Source	Probable Cause

c. Starting relay chatter.
d. Noisy motor bearings.
e. Single phasing a 3 phase motor.

10. SHAFT
    a. Bent.
    b. Undersized. May cause noise at impeller, bearings or sheave.
    c. If more than two bearings are on a shaft, they must be properly aligned.

11. HIGH AIR VELOCITY
    a. Duct work too small for application.
    b. Fan selection too small for application.
    c. Registers or grilles too small for application.
    d. Heating or cooling coil with insufficient face area for application.

12. OBSTRUCTION IN HIGH VELOCITY GAS STREAM MAY CAUSE RATTLE, OR PURE TONE WHISTLE
    a. Dampers.
    b. Registers.
    c. Grilles.
    d. Sharp elbows.
    e. Sudden expansion in duct work.
    f. Sudden contraction in duct work.
    g. Turning vanes.

13. PULSATION OR SURGE
    a. Restricted system causes fan to operate at poor point of rating.
    b. Fan too large for application.
    c. Ducts vibrate at same frequency as fan pulsations.

14. GAS VELOCITY THROUGH CRACKS, HOLES OR PAST OBSTRUCTIONS
    a. Leaks in duct work.
    b. Fins on coils.
    c. Registers or grilles.

15. RATTLES AND/OR RUMBLES
    a. Vibrating duct work.
    b. Vibrating cabinet parts.
    c. Vibrating parts not isolated from building.

B. LOW CFM, INSUFFICIENT AIR FLOW

1. FAN
    a. Forward curved impeller installed backwards.
    b. Fan running backwards.

B. LOW CFM, INSUFFICIENT AIR FLOW

Source	Probable Cause

	c. Cutoff missing or improperly installed.
	d. Impeller not centered with inlet collar(s).
	e. Fan speed too slow.
2. DUCT SYSTEM	a. Actual system is more restrictive (more resistance to flow) than expected.
	b. Dampers closed.
	c. Registers closed.
	d. Leaks in supply ducts.
	e. Insulating duct liner loose.
3. FILTERS	a. Dirty or clogged.
4. COILS	a. Dirty or clogged.
5. RECIRCULATION	a. Internal cabinet leaks in bulkhead separating fan outlet (pressure zone) from fan inlets (suction zone).
	b. Leaks around fan outlet at connection through cabinet bulkhead.
6. OBSTRUCTED FAN INLETS	a. Elbows, cabinet walls or other obstructions restrict air flow. Inlet obstructions cause more restrictive systems but do not cause increased negative pressure readings near the fan inlet(s). See Part 1 "System Effect." Fan speed may be increased to counteract the effect of restricted fan inlet(s).
7. NO STRAIGHT DUCT AT FAN OUTLET	a. Fans which are normally used in duct system are tested with a length of straight duct at the fan outlet. If there is no straight duct at the fan outlet, decreased performance will result. If it is not practical to install a straight section of duct at the fan outlet, the fan speed may be increased to overcome this pressure loss. See Part 1 "System Effect."
8. OBSTRUCTIONS IN HIGH VELOCITY AIR STREAM	a. Obstruction near fan outlet.
	b. Sharp elbows near fan outlet.
	c. Improperly designed turning vanes.

B. LOW CFM, INSUFFICIENT AIR FLOW

Source   **Probable Cause**

d. Projections, dampers or other obstruction in part of system where air velocity is high.

C. HIGH CFM, TOO MUCH AIR FLOW

1. SYSTEM

a. Oversized duct work.
b. Access door open.
c. Registers or grilles not installed.
d. Dampers set to by-pass coils.
e. Filter(s) not in place.

2. FAN

a. Backward inclined impeller installed backwards (HP will be high).
b. Fan speed too fast.

D. WRONG STATIC PRESSURE

1. SYSTEM, FAN OR INTERPRETATION OF MEASUREMENTS

GENERAL DISCUSSION
The velocity pressure at any point of measurement is a function of the velocity of the air or gas and its density.

The static pressure at a point of measurement in the system is a function of system design (resistance to flow), air density and the amount of air flowing through the system.

The static pressure measured in a "loose" or oversized system will be less than the static pressure in a "tight" or undersized system for the same airflow rate.

In most systems, pressure measurements are indicators of how the installation is operating. These measurements are the result of air flow and as such are useful indicators in defining system characteristics.

Field static pressure measurements rarely correspond with laboratory static pressure measurements unless the fan inlet and fan

D. STATIC PRESSURE LOW, CFM HIGH

Source	Probable Cause
	outlet conditions of the installation are exactly the same as the inlet and outlet conditions in the laboratory.
	Also see D-2 through D-6, E-2, F-1, and G-1, for specific cases.
2. SYSTEM	System has less resistance to flow than expected. This is a common occurrence. Fan speed may be reduced to obtain desired flow rate. This will reduce HP (operating cost).
3. GAS DENSITY	Pressures will be less with high temperature gases or at high altitudes.
4. FAN	a. Backward inclined impeller installed backwards. HP will be high. b. Fan speed too high.

E. STATIC PRESSURE LOW, CFM LOW

1. SYSTEM	a. Fan inlet and/or outlet conditions not same as tested. See general discussion (D.1). Also see Part 1 System Effect Factors.
	Also see B.1–8.

F. STATIC PRESSURE HIGH, CFM LOW

2. SYSTEM	a. Obstruction in system. b. Dirty filters. c. Dirty coil. d. System too restricted.
	Also see B.1–8.

G. HORSEPOWER HIGH

1. FAN	a. Backward inclined impeller installed backwards. b. Fan speed too high.
2. SYSTEM	a. Oversized duct work.

G. HORSEPOWER HIGH

Source	Probable Cause

b. Face and by-pass dampers oriented so coil dampers are open at same time by-pass dampers are open.

c. Filter(s) left out.

d. Access door open.

3. GAS DENSITY
    a. Calculated horsepower requirements based on light gas (e.g. high temperature) but actual gas is heavy (e.g. cold start up).

4. FAN SELECTION
    a. Fan not operating at efficient point of rating. Fan size or type may not be best for application.

H. FAN DOES NOT OPERATE

1. ELECTRICAL OR MECHANICAL

Mechanical and electrical problems are usually straightforward and are normally analyzed in a routine manner by service personnel. In this category are such items as:

a. Blown fuses.

b. Broken belts.

c. Loose pulleys.

d. Electricity turned off.

e. Impeller touching scroll.

f. Wrong voltage.

g. Motor too small and overload protector has broken circuit.

I. PREMATURE FAILURE

1. BELTS, BEARINGS, SHEAVES IMPELLER, HUBS, ETC.

GENERAL DISCUSSION

Each fan component is designed to operate satisfactorily for a reasonable life time. Fans intended for heavy duty service are made especially for that type of service. For example, Class I fans are intended for operation below certain limits of pressure and outlet velocity. Class II fans are designed for higher operating limits (see

I. PREMATURE FAILURE

Source	Probable Cause
	AMCA Standard 2408). Not all components are limited by the same factors, e.g. limiting factors may be HP, RPM, temperature, impeller tip speed, torque, corrosive atmospheres, expected life, etc. Also see A.3, A.5, A.6.
2. COUPLINGS	See A.4.
3. SHAFT	Also see A.10.

ROTATION

No system should be tested without first checking the fan rotation. Fan rotation, as defined by the fan manufacturer, is either the "clockwise" or "counterclockwise" spin of the fan impeller. But the rotation would depend upon the position of the viewer relative to the fan. When checking rotation for *centrifugal fans* the fan must be viewed from the drive side while it is

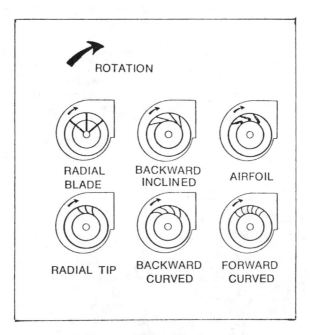

Figure 50A. Centrifugal Fan Impellers.

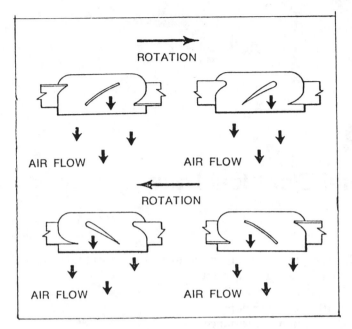

Figure 50B. Axial Fan Impellers.

coasting to a stop, see AMCA Standard 2406. For *tubular centrifugals*, the fan must be viewed from the outlet side. For *axial fans* the fans must be viewed from the inlet side.

Figures 50A and 50B show the correct rotation for centrifugal and axial fan impellers.

18
Useful Electrical Data

The following symbols are common usage in electrical formulas:

    I = Ampere, a unit of current

    E = Volt, a unit of pressure

   W = Watt, a measure of power

    R = Ohm, a unit of resistance

  eff = Efficiency; use 0.85 when eff is unknown

   pf = Power factor; the ratio of the actual power to the apparent power

 kw = 1000 watts, a measure of power

 hp = horsepower; 1 hp = 0.746 kw

Figure 51 shows the Power Equation Wheel from which the basic electrical formulas may be read. W may be substituted for P (power). Symbols in the outer ring equal the symbol in the inner ring.

To Find the Power Factor:

1. Single phase circuits; $pf = \dfrac{W}{E \times I}$

2. Two phase circuits; $pf = \dfrac{W}{E \times I \times 2}$

3. Three phase circuits; $pf = \dfrac{W}{E \times I \times 1.73}$

To Find Amps Where Hp is Known:

1. Single phase circuits; $I = \dfrac{hp \times 746}{E \times eff \times pf}$

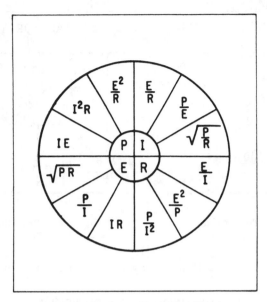

Figure 51. Power Equation Wheel.

2. Three phase circuits; $I = \dfrac{hp \times 746}{1.73 \times E \times eff \times pf}$

To Find HP where Amps are Known:

1. Single phase circuit; $hp = \dfrac{I \times E \times eff \times pf}{746}$

2. Three phase circuit; $hp = \dfrac{I \times E \times eff \times pf \times 1.73}{746}$

THREE PHASE AC CIRCUITS

Three phase is the most common polyphase system. It is connected in either "delta" or "star" (sometimes called Y) formation. Figure 52 shows a *delta* system, the voltage between any two wires is 240 V; the line voltage in a delta connection equals the coil voltage. Figure 53 shows a *star* system, the line voltage in a star connection equals the coil voltage $\times \sqrt{3}$, or 120 V $\times$ 1.732 = 208 V.

ELECTRICAL MEASUREMENTS

Electrical units of measurement are given in Table 18 together with the instruments generally used to measure each unit. Frequency, inductance

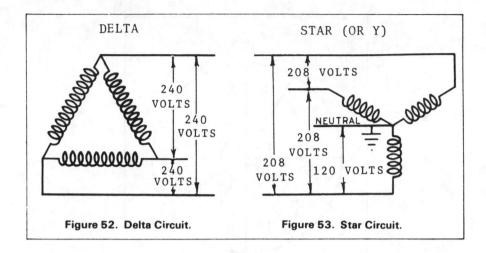

Figure 52. Delta Circuit.   Figure 53. Star Circuit.

(henry), and capacitance (farad) may also be measured but will not be
discussed here.

In Figure 54 the application for measuring instruments is shown graphically
in the circuit. Figure 54A shows a technician taking a reading with an
ammeter. Figure 54B illustrates the use of a tachometer for reading rpm.

As a matter of convenience, a prefix system is used to express very large or
very small quantities of electrical units, thereby avoiding the use of many
zeroes and decimals. Some common prefixes and their equivalents are:

Kilo = Thousands   1 kilowatt = 1000 watts
Meg   = Millions   1 megohm   = 1,000,000 ohms
Mili = 1 / 1000   1 miliamp   = 0.001 amp
Micro = 1 / 1,000,000   1 microamp = 0.000001 amp

MOTOR STARTERS

Figure 55 shows a typical line starter diagram. Normally the "start" button
contacts are open and no current passes through the starter. When the "start"

Table 18. Practical Absolute Units Of Electrical Quantities.

CODE	DESCRIPTION	SYMBOL	UNIT	MEASURING INSTRUMENT
Voltage (emf)	Pressure	V	Volt	Voltmeter
Resistance	Resistance to Flow	Ω	Ohm	Ohmeter
Current	Flow	I	Ampere	Ammeter
Power	Rate of Energy	W	Watt	Wattmeter
Power Factor	Power Ratio	pf		Power Factor Meter

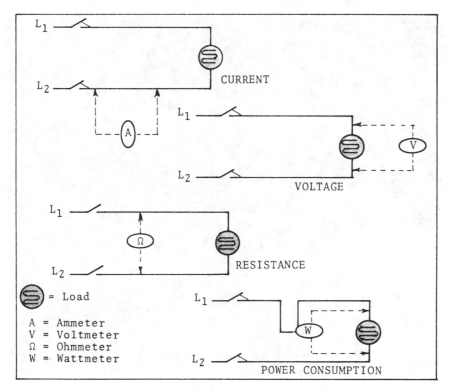

Figure 54. Electrical Measuring Instrument Hook-up.

button is depressed, current passes through the contactor, or magnetic coil. When the coil is energized, the plunger "P" is drawn up, magnetically raising the contact bar and completing the circuit between line and fan motor. When the coil is *de-energized*, the plunger, released from its magnetic field, drops by gravity and takes the contactor bar with it, thereby breaking the circuit. Releasing the "start" button does not break the circuit. Once the holding coil is energized, the only way to break the circuit is to depress the "stop" button.

When the temperature rise of a motor is excessive, the insulation life is shortened; this is the most serious effect of motor overload. To protect motors against burnout by excessive heat, the motor starter usually is equipped with *overload protection*. The most common device is the thermal bimetallic overload relay installed in the circuit adjacent to the heater coils through which the motor current passes. When abnormally high current passes through the heater coils, the overload relays will warp and interrupt the control circuit to the holding coil. See Figure 55.

The technician must check the rating of the installed heater elements against the recommended rating number usually found on the inside cover of the

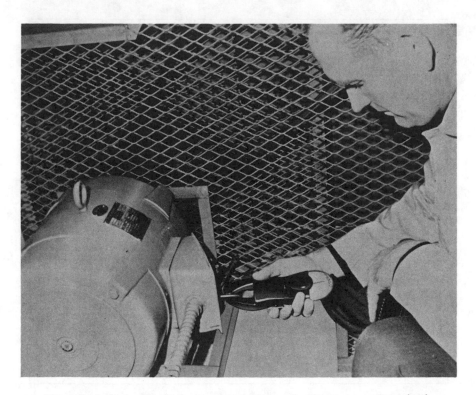

Figure 54A. Using the ammeter to read amps and volts across a motor lead.

starter. It is not unusual to find improperly sized heater elements in motor starters.

Upon startup, it may be found that the fan motor is rotating in the wrong direction. To reverse a three-phase machine it is only necessary to interchange two of the incoming leads. In the case of single-phase motors, it is necessary to reverse the starting winding relative to the running winding, therefore any links between the starting and running terminals should be removed.

Motors must, of course, be matched to the power supply. Ordinarily, the motor *nameplate* gives the name of the manufacturer, rating in horsepower, power supply frequency and voltage, full-load amperes, locked-rotor, temperature rise, service factor, duty cycle, and manufacturer's identification number.

TYPES OF ENCLOSURES

Surrounding atmospheric conditions determine the selection of motors. The more enclosed a motor, the more it costs and the hotter it runs.

Open-type: Allows maxium ventilation, has full opening in the frame and end bells.

Semi-protected: Openings are lined with screens to protect the motor from falling particles.

Drip-proof: The upper parts of the motor are shielded to prevent vertical overhead dripping.

Splash-proof: The bottom parts of the motor are shielded to prevent splashing or particles at an angle not over 100 degrees from vertical.

Totally enclosed: Nonventilated, may be used in hazardous atmospheres, or may be explosion-proof.

GLOSSARY OF ELECTRICAL TERMS

Ampere(I): Unit of electrical current produced in a conductor by the applied voltage. One ampere is equal to the flow of one coulomb passing a point in a circuit every second.

Brake Horsepower (bhp): The horsepower *actually* required to drive a fan; it includes the energy losses in the fan but does not include the drive loss between the motor and the fan. The name derives from the Prony brake, a common method of testing mechanical output of motors.

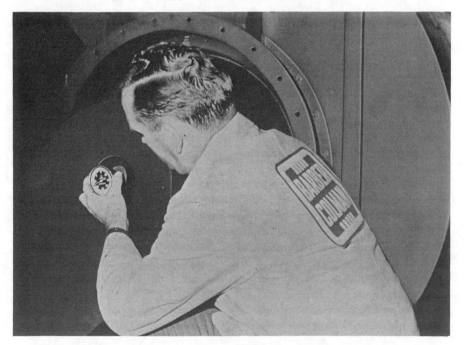

Figure 54B. Using a tachometer to check fan motor revolutions (rpm).

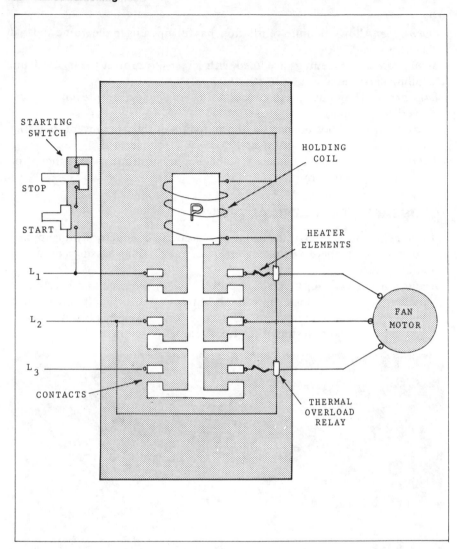

Figure 55. Motor Starter.

Coulomb (Q): The basic quantity of electrical measurement. One coulomb
   equals 6.25 billion billion electrons or one ampere per sec.

Electromotive Force (EMF): The force that causes current to flow, i.e., the
   electric potential difference between the terminals of any device used as a
   source of electrical energy.

Frequency: The number of complete cycles of alternating current per second.

Horsepower (hp): Work performed equivalent to 33,000 ft-lb per min. The term horsepower was first used by James Watt to measure the power of his steam engine against the power of a London draught horse.

Motor Efficiency (EFF): The rate at which a motor converts electrical energy. The greater the efficiency, the lower the energy cost. A well-designed induction motor may have an EFF between 75% and 95% full load.

Ohm (R): A unit of resistance to the flow of current.

Power Factor (PF): The ratio of actual to apparent power of an alternating current found by measuring the current with a wattmeter as against the indicated voltmeter reading. It is a measure of the loss in an insulator, capacitor, or inductor.

Service Factor: A safety factor designed into a motor to deliver more than its rated horsepower. A 10 hp motor with a service factor of 1.10 is capable of delivering 11 hp.

Temperature Rise: Part of the electrical specification of a motor; a measure of the motor's capacity to dissipate the heat generated by its electrical power losses. If the nameplate shows a temperature rise of 40 C and the ambient temperature is 90 F (32.2 C), the motor could operate at 162 F (72 C). If the motor becomes hot, hang one thermometer (0–110 C or 0–230 F) in the room near the motor, and place one thermometer in firm contact with the bearing bushing, stator, or other *hot* stationary part of the motor, and check the rise.

Torque: A rotating force. *Starting torque*—also known as locked-rotor or breakaway torque—is the torque available to start a load from a standstill i.e., the torque exerted by the starting current of the motor to overcome the static friction at rest.

Volt (V or E): A unit of electrical pressure that will push one ampere through a resistance of one ohm.

Watt (W): A unit of power being the amount of energy expended for one ampere to flow through one ohm.

19

Cost Estimating For Testing And Balancing

It is essential that the testing and balancing agency or department keep careful cost records and time cards from which to build a "task times information bank." There is no substitute for experiential costing records as a basis from which to develop in-house cost estimating procedures.

The independent test and balance agency will early learn that the man-hour requirements will be largely affected by variables such as; geography, different installing contractors, equipment manufacturers, and job conditions—and not least, by the architect and mechanical engineer. But average jobs do average out. The following tables suggest a basic man-hour reference for cost estimating.[1] To estimate the man-hour requirements for the various task times for testing, adjusting, and balancing, Tables 19 through 30 give the breakdown for the most common tasks. Figure 56 may be used for rule-of-thumb speed estimating.

Table 19. Air Handlers, Air Side.

OPERATION	MAN-HOURS
1. Record name plate information; remove panel and check fan rotation, check fan inclination, check number of rows of coil; replace panel; read rpm, amperes, and volts; adjust rpm; drill holes for thermometer traverse at coil entering and leaving positions; check entering and leaving dry-bulb and wet-bulb	3.0

[1]These tables were developed in 1966 and first appeared in *Mechanical Estimating Guidebook* 4th ed. by John Gladstone McGraw-Hill, New York, Copyright © 1970. Used by permission of McGraw-Hill Book Company.

Table 19. (continued)

OPERATION	MAN-HOURS
2. Discharge air: drill holes for duct traverse; set up instrument; make pitot tube traverse, read manometer, record velocity pressures and velocities; adjust dampers; second traverse and readout	1.7
3. Return air (if apparatus has return-air duct, use same number of man-hours as above in step 2) plenum: check air filters and filter pressure drop; measure free area of opening; make three anemometer traverses in opposing directions and average out; record velocities	1.0
4. Outside air: same as discharge air (step 2), but much smaller traverse area	0.8
5. Final: after all outlets have been checked and balanced, take final readings at existing traverse holes (above); make final adjustments; make final entries on report sheets; plug and seal all traverse holes	1.6
Total man-hours per air handler	8.1

For multizone units, add 1.3 man-hours per zone.

Table 20. Air Handlers, Water Side.

OPERATION	MAN-HOURS
After total water has been set at chiller: Install calibrated gages at coil; set pressure drop across bypass valve to match full drop at coil; read pressure drop and adjust to design; check and record gpm; make second adjustment; tag and mark all settings; remove gages; make final entries on report sheets	2.5
Note: If air handler has a venturi flow valve installed at time of erection, and balancing technician uses portable venturi flowmeter, reduce man-hours by 30%.	

Table 21. Chiller Apparatus.

OPERATION	MAN-HOURS
1. At each pump: Record nameplate information; check and record running amperes, voltage, rpm, and starter heater elements; check and record pressures and gpm	1.5
2. At each chiller: Set up manometer and pitot tube; make pitot tube traverse, check and record pressure drop and gpm; adjust to design; tag and mark all settings; remove instruments; make final entries on report sheets	2.8
Total man-hours	4.3

Note: Based on 2 in. nipple and gate valve, plus gage cocks, installed at suitable point in system at time of erection. If venturi flow valve has been installed at time of erection, and balancing technician uses portiable venturi flowmeter, reduce man-hours by 30%.

Table 22. Air Diffusers, Standard Ceiling Type.

OPERATION	MAN-HOURS
Move ladder and instruments into position; take four-point velometer reading, record velocity and cfm; adjust volume to design; take second round of readings; make final entries on report sheets	0.5

Note: If device handles less than 200 cfm, add 0.2 man-hour; if device handles more than 2,300 cfm, add 0.3 man-hour.

Table 23. Side-wall Grilles and Registers.

OPERATION	MAN-HOURS
Move ladder and instruments into position; take two 1-min anemometer traverses across face of grille, record velocity and cfm; adjust volume to design; take second round of readings; make final entries on report sheet	0.4

Note: If device handles less than 200 cfm, add 0.15 man-hour; if device handles more than 2,300 cfm, add 0.20 man-hour.

Table 24. Slim-line Linear Slot Diffusers.

OPERATION	MAN-HOURS
Move ladder and instruments into position; take velometer reading, record velocity and cfm; adjust volume to design; take second round of readings; make final entries on report sheets	0.8

Time unit based on per 10 ft of continuous diffuser.

Table 25. Miscellaneous Ventilation.

OPERATION	MAN-HOURS
Using proper instrumentation, test, balance, and make final entries on report sheets, including all nameplate data, etc.	
1. Small blowers, to 5,000 cfm	1.25
2. Large blowers, to 30,000 cfm	3.25
3. Roof exhausters	1.00
4. Range hoods	1.25
5. Laboratory hoods	1.50

Table 26. Combustion Test.

OPERATION	MAN-HOURS
Drill hole in smokestack and fire door; take CO_2 reading; check flue-gas temperature; check and adjust stack draft; make smoke test and adjust air shutter; adjust burner; record analysis; make final entries on report sheet	2

Table 27. Traversing Rooms with Sling Psychrometer.

OPERATION	MAN-HOURS
Pace off room and determine traverse points; wet wick and whirl psychrometer; record dry-bulb and wet-bulb; read relative humidity on chart or slide rule; make final entries on report sheet.	
1. Small rooms, per each 500 sq ft traversed	1
2. Large rooms, per each 1,000 sq ft traversed	1
3. Minimum allowance per room	0.33

Table 28. Test and Adjust Controls.

OPERATION	MAN-HOURS
Check for tight contact at all terminals; test for cycle in all positions; make necessary adjustments; make entries on report sheet	
1. Thermostat, two-position	0.40
2. Thermostat, modulating	0.80
3. Humidistat, two-position	0.80
4. Humidistat, modulating	1.25
5. Motor control	0.80
6. Water solenoid valve	0.30
7. Expansion valve	0.80
8. Back-pressure control	1.10
9. Water-regulating valve	1.00
10. Auto pump-down	1.25

Table 29. Hydrostatic Test.

OPERATION	MAN-HOURS
Hydrostatic test of erected piping system with cone-lock test plug: Prepare for test; set gage, check and record pressure; retest (after leaks have been repaired); make final entries on test report sheet	2.5
Note: Does not include filling time or standby.	

Table 30. Induction and Fan-coil Units.

OPERATION	MAN-HOURS
Based on floor-mounted or near-floor-mounted vertical cabinets for high-velocity induction units and fan-coil units:	
Test for leaks	0.25
Balance and adjust	1.00
Total man-hours	1.25
Note: For horizontal ceiling-mounted units add 0.30 man-hour.	

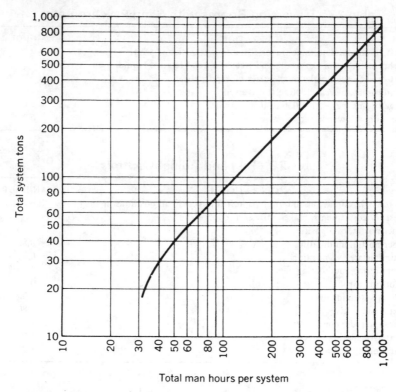

Total system tons

Total man hours per system

Figure 56. Speed-Sheet Estimating—System Testing And Balancing, Man-hours Per System.

20
Système International d'Unités Guide (Metrics)

Table 31. Most Commonly Used HPACV Units.

UNIT OF MEASURE	MULTIPLY	BY	TO OBTAIN	SYMBOL
DUCTWORK: AIR FLOW	Cfm	0.0004719	cubic metres/second	m^3/s
	Fpm	0.00508	metres/second	m/s
	Fpm	0.508	centimetres/second	cm/s
	Fps	0.3048	metres/second	m/s
	Mph	0.447	metres/second	m/s
DUCTWORK: PRESSURE	Inch H_2O	0.25	kilopascals	kPa
	Inch H_2O	249 (use 250)	pascals	Pa
	Inch H_2O/100′	8.176	pascals (N/m^2)	Pa
	Inch Hg	3.386	kilopascals (kN/m^2)	kPa
DUCTWORK: LENGTH & AREA	Inch	25.4	millimetres	mm
	Inch	2.54	centimetres	cm
	Inch	0.0254	metres	m
	In.^{2}	6.452	centimetres squared	cm^2
	Feet	0.3048	metres	m
	Ft^2	0.0929	metres squared	m^2
	Ft^3	0.02832	metres cubed	m^3
FAN DUTY	Cfm	0.4719	litres/second	l/s
	Inch H_2O	249 (use 250)	pascals	Pa
	Hp	0.7460	kilowatts	kW
	Rpm	0.10472	radians/sec	rad/s
	Rpm	60	revolutions/sec	rev/s
PUMP DUTY	Psf	47.88 (use 50)	pascals (N/m^2)	Pa
	Psi	6895 (use 7000)	pascals (N/m^2)	Pa
	Psi	6.895	kilopascals (kN/m^2)	kPa
	TO OBTAIN ABOVE	DIVIDE BY ABOVE	STARTING WITH ABOVE	

Table 31. (continued)

UNIT OF MEASURE	MULTIPLY	BY	TO OBTAIN	SYMBOL
	Gpm	0.00006309	metres cubed/second	m^3/s
	Gpm	0.06309	litres/second	l/s
ENERGY,	Btuh	0.2931	watts	W
WORK,	Btuh	0.0002931	kilowatt	kW
HEAT	Ton (ref)	3516	kilowatt	kW
	Ton	3.516	kilowatt	kW
	Btu	1054	joule (watt-sec)	J
	kWh	3.6	megajoule	MJ
THERMAL	Btu	1055.06	joule	J
FLOW	Btu	1.05506	kilojoule	kJ
	Btu/ft^3 deg F	67066	joule/metre³ deg C	J/m^3 deg C
	Btu/lb deg F	4186.8	joule/kilogram deg C	J/kg deg C
	Btuh	0.2931	watts	W
	Btu/ft^2 hr	3.155	watts/metre sq	W/m^2
	Btu/ft^2 deg F (U)	5.678	watts/metre sq deg C	W/m^2 deg C
	Temperature diff. F	0.555	temperature diff., C	ΔtC
	TO OBTAIN ABOVE	DIVIDE BY ABOVE	STARTING WITH ABOVE	

MORE PRESSURE CONVERSION FACTORS

Psi × 703.1 = mm H_2O
Psi × 51.71 = mm Hg
Psi × 0.0690 = bar
Psi × 68.95 = m bar
Psi × 0.06805 = atmosphere
Bar × 1 × 10^5 = N/m^2 (Pa)
Atmosphere × 101300 = N/m^2 (Pa)

Notes: 1. One Pascal = 1 Newton per metre squared N/m^2.
2. All operating pressures should be stated clearly in terms of gage pressure, e.g., "Gage pressure, 6 bar."

Air Flow, US

1. $Q = AV$

2. $V = 1096 \sqrt{\dfrac{V_p}{d}}$

   for standard air:

3. $V = 4005 \sqrt{V_p}$

4. $V_p = \left(\dfrac{V}{4005}\right)^2$

5. $V = V_m \sqrt{\dfrac{0.075}{d}}$ (other than standard air)

6. $T_p = V_p + S_p$

Q = Quantity, cfm
A = Area, ft^2
V = Velocity, fpm
V_p = Velocity pressure, in. WG
d = Air density, lb/ft^3
V_m = Measured velocity, fpm

T_p = Total pressure, in. WG
V_p = Velocity pressure, in. WG
S_p = Static pressure, in. WG

Fan Laws, US

7. $cfm_2 = cfm_1 \times \left(\dfrac{rpm_2}{rpm_1}\right)$

8. $Sp_2 = Sp_1 \times \left(\dfrac{rpm_2}{rpm_1}\right)^2$

9. $hp_2 = hp_1 \times \left(\dfrac{rpm_2}{rpm_1}\right)^3$

cfm = Air quantity, ft^3/min

rpm = Revolutions/min
Sp = Static pressure, in WG
hp = Horsepower

Pulley Laws, US

10. $Rpm,D = rpm,d \times \dfrac{diameter,d}{Diameter,D}$

11. $rpm,d = Rpm,D \times \dfrac{Diameter,D}{diameter,d}$

12. $Diameter,D = diameter,d \times \dfrac{rpm,d}{Rpm,d}$

Rpm,D = Revolutions/min, driven pulley

rpm,d = Revolutions/min, driver pulley

$Diameter,D$ = Diameter driven pulley, in.

Air Flow, SI

1. $Q = AV$

2. $V = 1.3 \sqrt{V_p\left(\dfrac{1.2}{d}\right)}$

   for standard air:

3. $V = 1.3 \sqrt{V_p}$

4. $V_p = \left(\dfrac{V}{1.3}\right)^2$

5. $V = V_m \sqrt{\dfrac{1.2}{d}}$ (other than standard air)

6. $T_p = V_p + S_p$

Q = Quantity, m^3/s
A = Area, m^2
V = Velocity, m/s
V_p = Velocity pressure, Pa
d = Air density, kg/m^3
V_m = Measured velocity, m/s

T_p = Total pressure, Pa
V_p = Velocity pressure, Pa
S_p = Static pressure, Pa

Fan Laws, SI

7. $m^3/s_2 = m^3/s_1 \times \left(\dfrac{rps_2}{rps_1}\right)$

8. $S_{p_2} = S_{p_1} \times \left(\dfrac{rps_2}{rps_1}\right)^2$

9. $kW_2 = kW_1 \times \left(\dfrac{rps_2}{rps_1}\right)^3$

m^3/s = Air quantity

rps = Revolutions/sec
S_p = Static pressure, Pa
kW = Kilowatts

Pulley Laws, SI

10. $Rps, D = rps, d \times \dfrac{diameter, d}{Diameter, D}$

11. $rps, d = Rps, D \times \dfrac{Diameter, D}{Diameter, d}$

12. $Diameter, D = diameter, d \times \dfrac{rps, d}{Rps, D}$

Rps,D = Revolutions/sec driven pulley
rps,d = Revolutions/sec, driver pulley

Diameter,D = Diameter driven pulley,
metres

13. $\text{diameter,d} = \text{Diameter,D} \times \dfrac{\text{Rpm,D}}{\text{rpm,d}}$ diameter,d = Diameter driver
pulley, in.

Gas Laws, US

14. $V_2 = V_1 \times \dfrac{P_1}{P_2}$

15. $V_2 = V_1 \times \dfrac{T_2}{T_1}$

16. $P_2 = P_1 \times \dfrac{T_2}{T_1}$

V_1 = Initial volume, ft^3
V_2 = Final volume, ft^3
T_1 = Initial temperature, absolute, °F
T_2 = Final temperature, absolute, °F
P_1 = Initial pressure, psia
P_2 = Final pressure, psia
Absolute temperature = °F + 460

Heat Transfer, US

Air:

17. $H = 60 \times C \times d \times cfm \times \Delta t$
simplified for standard air:
18. H, sensible = $1.08 \times cfm \times \Delta t$
19. H, latent = $.68 \times cfm \times \Delta gr/lb$
20. H, total = $4.5 \times cfm \times \Delta h$

H = Heat flow, Btuh
C = Specific heat, Btu/lb. °F
Δt = Temperature difference, °F
$\Delta gr/lb$ = Humidity ratio, grains/lb dry air.
Δh = Enthalpy difference, Btu/lb dry air

Water:

21. $H = 60 \times 8.33 \times Q \times \Delta t$
simplified for standard water:
22. $H = 500 \times Q \times \Delta t$

Q = Water flow,
gallons per minute, gpm
8.33 = Weight of 1 gallon of water
Δt = Temperature difference
between inlet and outlet water, °F

Solid Material:

23. $H = A \times U \times \Delta t$

H = Heat flow, Btuh
A = Area, ft^2

13. $\text{diameter}, d = \text{Diameter}, D \times \dfrac{\text{Rps}, D}{\text{rps}, d}$

diameter,d = Diameter driver pulley, metres

Gas Laws, SI

14. $V_2 = V_1 \times \dfrac{P_1}{P_2}$

15. $V_2 = V_1 \times \dfrac{T_2}{T_1}$

16. $P_2 = P_1 \times \dfrac{T_2}{T_1}$

V_1 = Initial volume, m^3
V_2 = Final volume, m^3
T_1 = Initial temperature, absolute, K
T_2 = Final temperature, absolute, K
P_1 = Initial absolute pressure, Pa
P_2 = Final absolute pressure, Pa

Heat Transfer, SI

Air:
17. $H = 3.6 \times C \times d \times \ell/s \times \Delta t$
    simplified for standard air;
18. H, sensible $= 1.23 \times \ell/s \times \Delta t$
19. H, latent $= 3 \times \ell/s \times \Delta W$
20. H, total $= 1.2 \times \ell/s \times \Delta h$

H = Heat flow, watts

C = Specific heat, W/kg · °C
Δt = Temperature difference, °C
ΔW = Humidity ratio, gH_2O/kg dry air
Δh = Enthalpy difference, kJ/kg dry air
ℓ/s = Air flow rate, litres per second

Water:
21. $H = 1000 \times 4.2 \times Q \times \Delta t$
    simplified for standard water:
22. $H = 4200 \times Q \times \Delta t$

Q = Water flow, ℓ/s
4.2 = Specific heat of water, kJ/kg · K
1000 = Density of water, kg/m^3
Δt = Temperature difference between inlet and outlet water, °C

Solid Material:
23. $H = A \times U \times \Delta t$

H = Heat flow, watts
A = Area, m^2

$$U = \text{Overall heat transfer coefficient,}$$
$$\text{Btu/ft}^2 \cdot \text{hr} \cdot {}^\circ\text{F}$$
$$\Delta t = \text{Temperature difference of the two}$$
$$\text{sides, } {}^\circ\text{F}$$

Fan & Pump Duty, US

Air:

24. $hp = \dfrac{QP}{6356 \times eff}$

where the efficiency is not known assume 62.8% or simplified:

25. $hp = \dfrac{QP}{4000}$

Water:

26. $hp = \dfrac{Qh}{3960 \times eff}$

where the efficiency is not known assume 70% or simplified;

27. $hp = \dfrac{Qh}{2800}$

28. $\Delta p = \left(\dfrac{Q}{C_v}\right)^2$

29. $Q = C_v \sqrt{\Delta p}$

$hp = $ Fan horsepower
$Q = $ Air quantity, cfm
$P = $ Fan pressure, in. WG
$6356 = $ a constant
$eff = $ Fan efficiency

$hp = $ Pump horsepower
$Q = $ Water flow, gpm
$h = $ Head, ft
$3960 = $ a constant
$eff = $ Pump efficiency

$\Delta p = $ Pressure difference, psi
$Q = $ Water flow, gpm
$C_v = $ Control valve constant

U = Overall heat transfer coefficient, $W/m^2 \cdot {}^\circ C$

Δt = Temperature difference of the two sides, $^\circ C$

Fan & Pump Duty, SI

Air:

24. $W = \dfrac{QP}{1000}$

25. See across

W = Fan power, watts
Q = Air quantity, ℓ/s
P = Fan pressure, Pa
1000 = a constant

Water:

26. $W = \dfrac{Qh}{1000}$

27. See across

28. $\Delta p = \left(\dfrac{Q}{C_v}\right)^2$

29. $Q = C_v \sqrt{\Delta p}$

W = Pump power, watts
Q = Water flow, ℓ/s
h = Head, Pa
1000 = a constant

Δp = Pressure difference, Pa
Q = Water flow, ℓ/s
C_v = Control valve constant

	US	SI
Density, standard air	$0.075 \ lb/ft^3$	$1.2 \ kg/m^3$
Density, water	$8.33 \ lb/gal$	$1000 \ kg/m^3$
Specific heat, standard air	$0.24 \ Btu/lb \cdot {}^\circ F$	$0.28 \ W/kg \cdot {}^\circ C$
Specific heat, water	$1 \ Btu/lb \cdot {}^\circ F$	$4190 \ J/kg \cdot K$

21

Basic Math Review for Technicians and Journeyman Mechanics

FRACTIONS

To Reduce Common Fractions: Divide the numerator and denominator by common divisors until further reduction is impossible.

$$63/81 = 3\overline{)63}^{\,21} \quad 3\overline{)81}^{\,27} = 21/27 = 7/9$$

To Reduce Improper Fractions: Divide the numerator by the denominator. the quotient being a whole number and the remainder the new numerator.

$$43/6 = 6\overline{)43} = 7\ 1/6$$

To Express a Fraction as a Decimal: Divide the numerator by the denominator-

$$3/4 = 4\overline{)3} = 4\overline{)3.00}^{\,0.75} = 0.75$$

To Reduce Complex Fractions: First express both numerator and denominator as simple fractions, then multiply the upper numerator by the lower denominator for the new numerator and the lower numerator by the upper denominator.

$$\frac{1\ 3/4}{5/6} = \frac{7/4}{5/6} = 42/20 = 20\overline{)42} = 2\ 1/10$$

To Reduce Fractions to a Common Denominator: Multiply the numerator of each fraction by the product of all of the denominators except its own for the new numerators and multiply all denominators together for the new common denominator.

$$2/3, 1/4, 3/5 = 40/60, 15/60, 36/60$$

To Add Fractions: Reduce to a common denominator and add the numerators.

$$3/4 + 2/3 = 9/12 + 8/12 = 17/12 = 1\ 5/12$$

To Subtract Fractions: Reduce to a common denominator and subtract the numerators.

$$3/4 - 2/3 = 9/12 - 8/12 = 1/12$$

To Multiply Fractions: Multiply the numerators for a new numerator and multiply the denominators for a new denominator.

$$3/4 \times 5/8 = 15/32$$

To Divide Fractions: Invert the divisor and multiply.

a. $3/4 \div 7/8 = 3/4 \times 8/7 = 24/28 = 6/7$

b. $1\ 1/8 \div 3/16 = 9/8 \div 3/16 = 9/8 \times 16/3 = 144/24 = 6/1 = 6$

DECIMALS

To Express a Decimal as a Fraction: Disregard the decimal point and write the figures as the numerator of the fraction. Write the denominator as 1 plus as many zeros after it as there were figures following the decimal.

$$0.0125 = \frac{125}{10,000}$$

To Express a Fraction as a Decimal: Divide the numerator by the denominator.

$$32/64 = 1/2 = 2\overline{)1.0} = 0.5$$

To Multiply Decimals: Proceed as in simple multiplication and point off, as many decimal places in the product as are in the multiplier and multiplicand together.

$$\$87.96 \times 23.5 = \qquad 87.96 = 3 \text{ places}$$

$$\begin{array}{r} 87.96 \\ \times\ 23.5 \\ \hline 43980 \\ 26388 \\ 17592 \\ \hline \$2067.060\ (3\ \text{places}) \end{array}$$

To Divide Decimals: Proceed as in simple division and point off, as many decimal places in the quotient as are in the dividend in excess of the divisor.

$$0.2546 \div 0.38 = .38\overline{)\underset{}{.25\ 46}} = 0.67$$

$$\begin{array}{r} .67 \\ 22\ 8 \\ \hline 26\ 6 \\ 26\ 6 \end{array}$$

RATIO AND PROPORTION

Ratio is the relation of one figure to another and is sometimes expressed as a fraction with the first quantity as the numerator:

The ratio of 1 to 2 = 1:2 = 1/2

When ratios are equal to each other they are said to be in proportion. The ratio of 3 to 6 = 3:6 = 1/2, therefore it is equal to and in proportion to the ratio of 1 to 2 and the proportion would be written,

$$3:6 = 1:2$$

and, read, "3 is to 6 as 1 is to 2."

The first and last terms in a statement of proportion are called the *extremes*, and the middle terms are the *means*. A rule of proportion is that the product of the extremes is equal to the product of the means; therefore, in the above example:

$$3:6 = 1:2 = 3 \times 2 = 6 \text{ and } 6 \times 1 = 6$$

When the middle terms are indentical this quantity is called the mean proportional of the first and last terms.

$1:2 = 2:4$, 2 is the mean proportional between 1 and 4. To find the mean proportional of any two terms multiply them and extract the square root of their product. Thus the mean proportional of 2 and 50 is

$$\sqrt{2 \times 50} = \sqrt{100} = 10. \text{ Therefore, } 2:10 = 10:50$$

If a proportion is expressed algebraically as

$$a:b = c:d$$

then given any three terms the fourth can be determined. In direct proportions we may "solve for X" by one of the following formulas:

1. $a = \dfrac{b\,c}{d}$ 3. $b = \dfrac{a\,d}{c}$

2. $c = \dfrac{a\,d}{b}$ 4. $d = \dfrac{b\,c}{a}$

In the following examples, solve for X by using the above formulas;

1. $24:4$ as $X:3 = c = \dfrac{a\,d}{b} = \dfrac{24 \times 3}{4} = \dfrac{72}{4} = 18 = 24:4$ as $18:3$

2. $2:3$ as $4:X = d = \dfrac{b\,c}{a}$

$= \dfrac{3 \times 4}{2} = \dfrac{12}{2} = 6 = 2:3$ as $4:6$

3. $5:X$ as $25:20 = b = \dfrac{a\,d}{c}$

$= \dfrac{5 \times 20}{25} = \dfrac{100}{25} = 4 = 5:4$ as $25:20$

4. $X:25$ as $10:2 = a = \dfrac{b\,c}{d}$

$= \dfrac{25 \times 10}{2} = \dfrac{250}{2} = 125 = 125:25$ as $10:2$

Direct proportion formulas are useful in solving many problems, for example,

5. A 20 ft wall, shades 3/4 of a lawn, how much higher must the wall be to shade the entire lawn?

$$20:3/4 \text{ as } X:1 = 20:0.75 \text{ as } X:1 =$$

$$\frac{20 \times 1}{0.75} = 26.6 \quad 26.6 - 20 = 6.6 \text{ ft}$$

SQUARES, CUBES, AND POWERS

The square of any number is that number multiplied by itself once, it is called the *second power*, or

$$\text{the square of } 4 = 4 \times 4 = 16$$

Expressed algebraically; $4^2 = 16$. The small number ^{2} is called a *superscript* or an exponent. It is placed to the right of, and above any number.

$$6^2 = 6 \times 6 = 36, \text{ or } 6 \text{ squared} = 36$$

The exponent ^{2} means that the number (in this case 6) shall be used as a factor twice. If the exponent is 6^3 it means that the number 6 shall be used as a factor three times; $6 \times 6 \times 6 = 216$. We may say that 6 raised to the third power is 216, or

$$\text{the cube of } 6 = 6 \times 6 \times 6 = 216$$

A number may be raised to any power. Often the exponent is used as a shorthand to avoid the use of many zeroes. For example thirty-six million may be written 36,000,000. Notice that this number contains 6 zeroes. It may be written algebraically, 36×10^6 which simply signals the reader to add six zeroes to the number 36; 144,000,000,000 may be expressed as 144×10^9.

Conversely, when a superscript is written with a minus sign in front of it as 10^{-3} it signals the reader to move three decimals out. So $10^{-3} = 0.001$, and $36 \times 10^{-3} = 0.036$, and $36 \times 10^{-2} = 0.36$.

The use of superscripts are very important for converting metric SI units. For example, to convert fpm to m/s (feet per minute to metres per second) the conversion factor is 5.080×10^{-3}. Therefore, to convert 900 fpm to m/s,

$$900 \times 5.080 \times 10^{-3} = 900 \times 0.00508 = 4.57 \text{ m/s}$$

ROOTS AND RADICALS

The root of a number is the opposite operation of raising it to the power. The square root of $16 = 4$. Expressed algebraically; $\sqrt{16} = 4$. The sign indicating square root $\sqrt{}$ is called a radical.

The cube root of a number is the opposite of the cube. The cube root of 216 is 6 and it is indicated by the radical $\sqrt[3]{}$, therefore:

$$\sqrt[3]{216} = 6$$

Table 35 in the Appendix gives squares, roots and cubes for numbers from 0.01 to 1000.

EQUATIONS

An equation is a mathematical formulation which states that two quantities or expressions are equal. When an equation groups two or more numbers together between parentheses it means the group is to be considered as a single quantity. Therefore, the operation enclosed between parentheses should be isolated and carried out first.

Brackets indicate that the quantity of parentheses enclosed are to be considered as a single quantity. First isolate the parenthesis and carry out its operation; then isolate the brackets and carry out its operation.

Example:

$$\frac{6[3(4+8)+9]}{10} = \frac{6[3(12)+9]}{10} = \frac{6[45]}{10} = \frac{6 \times 45}{10} = \frac{270}{10} = 27$$

Observe that had the operation been carried out seriatim, the solution would have read, $6 \times 3 \times 4 + 8 + 9/10 = 8.9$, which would have been incorrect.

Example:

   Given: Effective room sensible heat $= 192,255$ Btuh
          Room temperature $= 75$ F
          Apparatus dew point $= 48$ F
          Coil bypass factor $= 0.05$
  Find: The dehumidified air quantity, cfm.
Substituting for the equation,

$$cfm_{da} = \frac{ERSH}{1.08(1 - BF)(t_{rm} - t_{adp})},$$

$$\frac{192{,}255}{1.08 \times (1 - 0.05) \times (75 - 48)} = \frac{192{,}255}{1.08 \times (0.95) \times (27)}$$

$$= \frac{192{,}255}{27.7} \qquad\qquad = 6940 \text{ cfm}$$

Graphic Equations for maneuvering:

By covering up any piece of the pie, it is possible to solve for that piece;

$$B \times C = A \qquad \frac{A}{B} = C \qquad \frac{A}{C} = B \quad \text{Let, } A = 24, B = 6, C = 4$$

$$4 \times 6 = 24 \qquad \frac{24}{6} = 4 \qquad \frac{24}{4} = 6$$

Now, applying these graphics to the *Fan laws:* The three main fan laws are;

1. $\dfrac{cfm_2}{cfm_1} = \dfrac{rpm_2}{rpm_1}$ cfm varies as fan speed

2. $\dfrac{sp_2}{sp_1} = \left(\dfrac{rpm_2}{rpm_1}\right)^2$ pressure varies as square of fan speed

3. $\dfrac{hp_2}{hp_1} = \left(\dfrac{rpm_2}{rpm_1}\right)^3$ power varies as cube of fan speed

By covering up the quarter of the pie which is unknown and given the three remaining quantities known, it is possible to solve for the unknown quantity;

1. Covering cfm_2, the equation remains, $cfm_2 = cfm_1 \times \left(\dfrac{rpm_2}{rpm_1}\right)$

2. Covering Sp_2, the equation remains, $Sp_2 = Sp_1 \times \left(\dfrac{rpm_2}{rpm_1}\right)^2$

3. Covering hp_2, the equation remains, $hp_2 = hp_1 \times \left(\dfrac{rpm_2}{rpm_1}\right)^3$

SIGNED NUMBERS

Frequently, the test and balance technician needs to manipulate negative numbers. For example, fan pressure measurements in the field read:

Fan inlet, $Sp = -1.5$ in.
$\qquad\qquad Vp = +0.25$ in.
Fan outlet, $Sp = +0.75$ in.
$\qquad\qquad Vp = +0.50$ in.

How is the *fan static pressure* calculated from these data? Figure 26 shows the hook-up for measuring the fan static pressure (Sp), where,

$$\text{Fan Sp} = \text{outlet Sp} - \text{inlet Tp}$$

Substituting for this formula,

$$\text{Fan Sp} = +0.75 \text{ in.} - (-1.25) = 2 \text{ in.}$$

The inlet Tp (-1.25) is found by substitution,

$$\text{Fan inlet Tp} = \text{Sp} + \text{Vp} = -1.5 + 0.25 = -1.25, \text{ Tp}$$

Signed Numbers. To deal with numbers which have a value of something less than zero, "negative numbers," it is necessary to understand how to manipulate these *signed numbers*. In the example of fan static pressure, above, both addition and subtraction of signed numbers was required.

Negative numbers are written with a *minus sign* preceding. Numbers having a value above zero are written with a *plus sign*. These are called *signed numbers*. If no sign is indicated the plus sign must be assumed.

Addition of Signed Numbers. To add numbers of *like signs*, add the numbers and give the like sign.

Example:

>   Add, 2 + 2 + 9
> *Solution:* 2 + 2 + 9 = + 13
>   Add, −2 + (−8) + (−6)
> *Solution:* −2 + (−8) + (−6) = − 16

To add numbers of *unlike signs,* separate the positives from the negatives and form two groups. Then subtract the smaller from the larger and answer in the sign of the larger.

Example:

>   Add, −8 + 6 + (−7) + 4 + 16 + (−2)
> *Solution:* 6 + 4 + 16 = +26
>   −8 + (−7) + (−2) = − 17
>   26 − 17 = 9

Subtraction of Signed Numbers. To subtract signed numbers, change the sign of the number to be subtracted (subtrahend) and follow the rule as for adding.

Example:

>   From −50, subtract −30
> *Solution:* Change the sign of the subtrahend giving,
>   (−50) − (−30) = −50 + 30 = −20

Multiplication of Signed Numbers. The product of any two numbers of *like signs* will always be *positive*; whereas the product of any two numbers of *unlike signs* will always be *negative*.

Example: (like signs)

> 90 × 3 = 270
> (−90) × (−3) = +270

Example: (unlike signs)

> 1. (−70) × (+9) = −630
> 2. (−30) × (−8) × (+4) × (−20) = −19,200
>   a. (−30) × (−8) = +240
>   b. (+240) × (+4) = +960
>   c. (+960) × (−20) = −19,200

Division of Signed Numbers. The quotient of any two numbers of *like signs* will always be *positive;* whereas the quotient of any two numbers of *unlike signs* will always be *negative.*

Example: (like signs)

$-120/-4 = +30$

Example: (unlike signs)

$-120/+4 = -30$

Applying the above rules, convert -40 Fahrenheit to Celsius. From the conversion formula;

$$°C = °F - 32 \times 5/9$$

a. $(-40) - (+32) = -40$
$$\begin{array}{r} -32 \\ \hline = -72 \end{array}$$
b. $-72 \times 5/9 = (-72) \times 0.555 = -40$ C
Convert -29 Celsius to Fahrenheit,

$$F = (C \times 9/5) + 32$$

a. $-29 \times (1.80) = -52.2$
b. $-52.2 + 32 = -20$ F

Bibliography

Air Velocity Instruments, Bulletin H-100, Dwyer Instruments Co.

Alden, John and Kane, John, *Design of Industrial Exhaust Systems*, 4th ed. Industrial Press, New York, 1970.

ASHRAE Handbook of Fundamentals, 1977, Chapter 13.

ASHRAE Handbook, Systems, 1976, Chapter 40.

Balancing Manual, BM 9-70, Tuttle and Bailey, New Britain, Connecticut.

Benedict, Robert P., *Fundamentals of Temperature, Pressure and Flow Measurements*, John Wiley., New York, 1969.

Carlson, G. F., "Hydronic Systems: Analysis and Evaluation," *ASHRAE Journal* Oct., 1968; Nov., 1968; Dec., 1968; Jan., 1969; Feb., 1969; March, 1969 (six articles).

Donnelly, F. M., "Preventing Fires in Kitchen Ventilating Ducts," *Air Conditioning, Heating and Ventilating*, Dec., 1966.

Fan Application Manual, Air Moving and Conditioning Association, 1976.

Fan Engineering, Buffalo Forge, Buffalo, New York, 1970.

Fans and Their Application in Air Conditioning, The Trane Company, La Crosse, Wisconsin, 1970.

Giammer, R. D., Locklin, D. W., and Talbert, S. G., "Preliminary Study of Ventilation Requirements for Commercial Kitchen," *ASHRAE Journal*, Nov., 1971, pp. 51–55.

Gladstone, J., *Mechanical Estimating Guidebook*, 4th ed., McGraw-Hill, New York, 1970, Chapter 15.

Gladstone, J., "Testing and Balancing of Air Conditioning Systems: State of the Art," *Heating, Piping and Air Conditioning*, Feb., 1968, p. 112.

Gladstone, J., "Graphs Simplify Fan Calculations", *Heating, Piping and Air Conditioning*, September, 1977, pp. 94–104.

Haessler, W. M., "A New Approach to Range Hood Protection," Presented at the 71st Annual Meeting of the National Fire Protection Association, May 1967, Boston, Mass.

Hemeon, W. C. L., *Plant and Process Ventilation*, 2nd ed., Industrial Press, New York, 1963.

Industrial Ventilation, 13th ed., American Conference of Governmental Industrial Hygienists, 1974.

Kahoe, H. T., "Graphs Simplify Field Testing and Balancing," *Air Conditioning and Refrigeration Business*, Nov., 1967, p. 45.

Kahoe, H. T., "How to Test and Balance Air Conditioning Systems," *Air Conditioning and Refrigeration Business*, Jan., 1963; p. 39; May, 1963, p. 48; Aug., 1963, p. 41 (three articles).

Kahoe, H. T., "Wet Finger Balancing Has Got to Go," *Air Conditioning and Refrigeration Business*, Sept., 1972, p. 73.

Ma, W. Y. L., "The Averaging Pressure Tubes Flowmeter for the Measurement of the Rate of Airflow in Ventilating Ducts and for the Balancing of Airflow Circuits in Ventilating Systems," *IHVE Journal*, Feb., 1967.

Mannion, C. F., "Information Points: Key to Water System Balancing," *Heating, Piping and Air Conditioning*, Jan., 1967, p. 145.

Manual for the Balancing and Adjustment of Air Distribution Systems, SMACNA, 1967.

National Standards for Field Measurements and Instrumentation, AABC, 1967.

Procedural Standards for Testing, Adjusting, Balancing, of Environmental Systems, National Environmental Balancing Bureau, 1977.

Tuve, G. L., "Measuring Air Flow," *Heating, Piping and Air Conditioning*, Dec., 1941.

Trane Air Conditioning Manual, The Trane Co., 1965, Chapter 9.

Uni-Flow Air Distribution Balancing Bulletin, Barber Colman, Rockford, Illinois.

Wind, M., "Air Balancing and Testing," *Heating, Piping and Air Conditioning*, Dec., 1968, p. 112.

Appendix
(Tables and Reference Data)

Table 32. Useful Conversion Factors.

Multiply	By	To Obtain
Atmosphere	29.92	Inches of mercury
Atmosphere	33.93	Feet of water
Atmosphere	14.70	Pounds per sq in.
Atmosphere	1.058	Tons per square foot
Barrels (oil)	42	Gallons
Boiler horsepower	33,475	Btu per hour
Boiler horsepower	34.5	Pounds water evaporated from and at 212F
Btu	778	Foot-pounds
Btu	0.000393	Horsepower-hours
Btu	0.000293	Kilowatt-hours
Btu	0.0010307	Pounds water evaporated from and at 212F
Btu per 24 hr	0.00000347	Tons of refrigeration
Btu per hour	0.00002986	Boiler horsepower
Btu per hour	0.000393	Horsepower
Btu per hour	0.000293	Kilowatts
Btu per inch per sq ft per hr per F.	0.0833	Btu per foot per sq ft per hour per F
Cubic feet	1,728	Cubic inches
Cubic feet	7.48052	Gallons
Cubic feet of water	62.37	Pounds (at 60F)
Cubic feet per minute	0.1247	Gallons per second
Feet of water	0.881	Inches of mercury (at 32F)
Feet of water	62.37	Pounds per sq ft
Feet of water	0.4335	Pounds per sq in
Feet of water	0.02950	Atmospheres
Feet per minute	0.01136	Miles per hour
Feet per minute	0.01667	Feet per second
Foot-pounds	0.001286	Btu
Gallons (U.S.)	0.1337	Cubic feet
Gallons (U.S.)	231	Cubic inches
Gallons of water	8.3453	Pounds of water (at 60F)
Horsepower	550	Foot-pounds per sec
Horsepower	33,000	Foot-pounds per min
Horsepower	2,546	Btu per hour
Horsepower	42.42	Btu per minute
Horsepower	0.7457	Kilowatts
Horsepower (boiler)	33,475	Btu per hour

Table 32. (continued)

Multiply	By	To Obtain
Inches of mercury (at 62F)	13.57	In. of Water (at 62F)
Inches of mercury (at 62F)	1.131	Ft of water (at 62F)
Inches of mercury (at 62F)	70.73	Pounds per sq ft
Inches of mercury (at 62F)	0.4912	Pounds per sq in
Inches of water (at 62F)	0.07355	Inches of mercury
Inches of water (at 62F)	0.03613	Pounds per sq in
Inches of water (at 62F)	5.202	Pounds per sq ft
Inches of water (at 62F)	0.002458	Atmospheres
Kilowatts	56.92	Btu per minute
Kilowatts	1.341	Horsepower
Kilowatt-hours	3415	Btu
Latent heat of ice	143.33	Btu per pound
Pounds	7,000	Grains
Pounds of water (at 60F)	0.01602	Cubic feet
Pounds of water (at 60F)	27.68	Cubic inches
Pounds of water (at 60F)	0.1198	Gallons
Pounds of water evaporated from and at 212F	0.284	Kilowatt-hours
Pounds of water evaporated from and at 212F	0.381	Horsepower-hours
Pounds of water evaporated from and at 212F	970.4	Btu
Pounds per square inch	2.0416	In. of mercury
Pounds per square inch	2.309	Ft of water (at 62F)

Table 33. Decimals Of A Foot.

In Frac.	0"	1"	2"	3"	4"	5"	6"	7"	8"	9"	10"	11"
0	.000	.083	.167	.250	.333	.417	.500	.583	.667	.750	.833	.917
1/8"	.010	.094	.177	.260	.344	.427	.510	.594	.677	.760	.844	.927
1/4"	.021	.104	.188	.271	.354	.438	.521	.604	.688	.771	.854	.938
3/8"	.031	.115	.198	.281	.365	.448	.531	.615	.698	.781	.865	.948
1/2"	.042	.125	.208	.292	.375	.458	.542	.625	.708	.792	.875	.958
5/8"	.052	.135	.219	.302	.385	.469	.552	.635	.719	.802	.885	.969
3/4"	.063	.146	.229	.313	.396	.479	.563	.646	.729	.813	.896	.979
7/8"	.073	.156	.240	.323	.406	.490	.573	.656	.740	.823	.906	.990

To change decimals of a foot to inches, multiply the decimal by 12.
To change inches to decimals of a foot, divide inches by 12.

Table 34. Inch-Foot-Decimal Conversion.

Inches, Fractions	Inches, Decimals	Feet, Decimals	Inches, Fractions	Inches, Decimals	Feet, Decimals	Inches, Fractions	Inches, Decimals	Feet, Decimals
1/64	.0156	.0013	11/32	.3438	.0287	43/64	.6719	.0560
1/32	.0313	.0026	23/64	.3594	.0299	11/16	.6875	.0573
3/64	.0469	.0039				45/64	.7031	.0586
1/16	.0625	.0052	3/8	.3750	.0313	23/32	.7188	.0599
5/64	.0781	.0065	25/64	.3906	.0326	47/64	.7344	.0612
3/32	.0938	.0078	13/32	.4063	.0339			
7/64	.1094	.0091	27/64	.4219	.0352	3/4	.7500	.0625
			7/16	.4375	.0365	49/64	.7656	.0638
1/8	.1250	.0104	29/64	.4531	.0378	25/32	.7813	.0651
9/64	.1406	.0117	15/32	.4688	.0391	51/64	.7969	.0664
5/32	.1563	.0130	31/64	.4844	.0404	13/16	.8125	.0677
11/64	.1719	.0143				53/64	.8281	.0690
3/16	.1875	.0156	1/2	.5000	.0417	27/32	.8437	.0703
13/64	.2031	.0169	33/64	.5156	.0430	55/64	.8594	.0716
7/32	.2188	.0182	17/32	.5313	.0443			
15/64	.2343	.0195	35/64	.5469	.0456	7/8	.8750	.0729
			9/16	.5625	.0469	57/64	.8906	.0742
1/4	.2500	.0208	37/64	.5781	.0482	29/32	.9063	.0755
17/64	.2656	.0221	19/32	.5938	.0495	59/64	.9219	.0768
9/32	.2813	.0234	39/64	.6094	.0508	15/16	.9375	.0781
19/64	.2969	.0247	5/8	.6250	.0521	61/64	.9531	.0794
5/16	.3125	.0260	41/64	.6406	.0534	31/32	.9688	.0807
21/64	.3281	.0273	21/32	.6563	.0547	63/64	.9844	.0820

Table 35. Squares, Cubes, Square Roots, and Cube Roots.

No.	Square	Cube	Square Root	Cube Root	No.	Square	Cube	Square Root	Cube Root
.01	.0001	.000001	0.1000	0.2154	.50	.2500	.125000	0.7071	0.7937
.02	.0004	.000008	0.1414	0.2714	.51	.2601	.132651	0.7141	0.7990
.03	.0009	.000027	0.1732	0.3107	.52	.2704	.140608	0.7211	0.8041
.04	.0016	.000064	0.2000	0.3420	.53	.2809	.148877	0.7280	0.8093
.05	.0025	.000125	0.2236	0.3684	.54	.2916	.157464	0.7348	0.8143
.06	.0036	.000216	0.2449	0.3915	.55	.3025	.166375	0.7416	0.8193
.07	.0049	.000343	0.2646	0.4121	.56	.3136	.175616	0.7483	0.8243
.08	.0064	.000512	0.2828	0.4309	.57	.3249	.185193	0.7550	0.8291
.09	.0081	.000729	0.3000	0.4481	.58	.3364	.195112	0.7616	0.8340
.10	.0100	.001000	0.3162	0.4642	.59	.3481	.205379	0.7681	0.8387
.11	.0121	.001331	0.3317	0.4791	.60	.3600	.216000	0.7746	0.8434
.12	.0144	.001728	0.3464	0.4932	.61	.3721	.226981	0.7810	0.8481
.13	.0169	.002197	0.3606	0.5066	.62	.3844	.238328	0.7874	0.8527
.14	.0196	.002744	0.3742	0.5192	.63	.3969	.250047	0.7937	0.8573
.15	.0225	.003375	0.3873	0.5313	.64	.4096	.262144	0.8000	0.8618
.16	.0256	.004096	0.4000	0.5429	.65	.4225	.274625	0.8062	0.8662
.17	.0289	.004913	0.4123	0.5540	.66	.4356	.287496	0.8124	0.8707
.18	.0324	.005832	0.4243	0.5646	.67	.4489	.300763	0.8185	0.8750
.19	.0361	.006859	0.4359	0.5749	.68	.4624	.314432	0.8246	0.8794
.20	.0400	.008000	0.4472	0.5848	.69	.4761	.328509	0.8307	0.8837
.21	.0441	.009261	0.4583	0.5944	.70	.4900	.343000	0.8367	0.8879
.22	.0484	.010648	0.4690	0.6037	.71	.5041	.357911	0.8426	0.8921
.23	.0529	.012167	0.4796	0.6127	.72	.5184	.373248	0.8485	0.8963
.24	.0576	.013824	0.4899	0.6214	.73	.5329	.389017	0.8544	0.9004
.25	.0625	.015625	0.5000	0.6300	.74	.5476	.405224	0.8602	0.9045
.26	.0676	.017576	0.5099	0.6383	.75	.5625	.421875	0.8660	0.9086
.27	.0729	.019683	0.5196	0.6463	.76	.5776	.438976	0.8718	0.9126
.28	.0784	.021952	0.5292	0.6542	.77	.5929	.456533	0.8775	0.9166
.29	.0841	.024389	0.5385	0.6619	.78	.6084	.474552	0.8832	0.9205
.30	.0900	.027000	0.5477	0.6694	.79	.6241	.493039	0.8888	0.9244
.31	.0961	.029791	0.5568	0.6768	.80	.6400	.512000	0.8944	0.9283
.32	.1024	.032768	0.5657	0.6840	.81	.6561	.531441	0.9000	0.9322
.33	.1089	.035937	0.5745	0.6910	.82	.6724	.551368	0.9055	0.9360
.34	.1156	.039304	0.5831	0.6980	.83	.6889	.571787	0.9110	0.9398
.35	.1225	.042875	0.5916	0.7047	.84	.7056	.592704	0.9165	0.9435
.36	.1296	.046656	0.6000	0.7114	.85	.7225	.614125	0.9220	0.9473
.37	.1369	.050653	0.6083	0.7179	.86	.7396	.636056	0.9274	0.9510
.38	.1444	.054872	0.6164	0.7243	.87	.7569	.658503	0.9327	0.9546
.39	.1521	.059319	0.6245	0.7306	.88	.7744	.681472	0.9381	0.9583
.40	.1600	.064000	0.6325	0.7368	.89	.7921	.704969	0.9434	0.9619
.41	.1681	.068921	0.6403	0.7429	.90	.8100	.729000	0.9487	0.9655
.42	.1764	.074088	0.6481	0.7489	.91	.8281	.753571	0.9539	0.9691
.43	.1849	.079507	0.6557	0.7548	.92	.8464	.778688	0.9592	0.9726
.44	.1936	.085184	0.6633	0.7606	.93	.8649	.804357	0.9644	0.9761
.45	.2025	.091125	0.6708	0.7663	.94	.8836	.830584	0.9695	0.9796
.46	.2116	.097336	0.6782	0.7719	.95	.9025	.857375	0.9747	0.9830
.47	.2209	.103823	0.6856	0.7775	.96	.9216	.884736	0.9798	0.9865
.48	.2304	.110592	0.6928	0.7830	.97	.9409	.912673	0.9849	0.9899
.49	.2401	.117649	0.7000	0.7884	.98	.9604	.941192	0.9899	0.9933
					.99	.9801	.970299	0.9950	0.9967

No.	Square	Cube	Square Root	Cube Root		No.	Square	Cube	Square Root	Cube Root
1	1	1	1.0000	1.0000		50	2500	125000	7.0711	3.6840
2	4	8	1.4142	1.2599		51	2601	132651	7.1414	3.7084
3	9	27	1.7321	1.4422		52	2704	140608	7.2111	3.7325
4	16	64	2.0000	1.5874		53	2809	148877	7.2801	3.7563
5	25	125	2.2361	1.7100		54	2916	157464	7.3485	3.7798
6	36	216	2.4495	1.8171		55	3025	166375	7.4162	3.8030
7	49	343	2.6458	1.9129		56	3136	175616	7.4833	3.8259
8	64	512	2.8284	2.0000		57	3249	185193	7.5498	3.8485
9	81	729	3.0000	2.0801		58	3364	195112	7.6158	3.8709
10	100	1000	3.1623	2.1544		59	3481	205379	7.6811	3.8930
11	121	1331	3.3166	2.2240		60	3600	216000	7.7460	3.9149
12	144	1728	3.4641	2.2894		61	3721	226981	7.8102	3.9365
13	169	2197	3.6056	2.3513		62	3844	238328	7.8740	3.9579
14	196	2744	3.7417	2.4101		63	3969	250047	7.9373	3.9791
15	225	3375	3.8730	2.4662		64	4096	262144	8.0000	4.0000
16	256	4096	4.0000	2.5198		65	4225	274625	8.0623	4.0207
17	289	4913	4.1231	2.5713		66	4356	287496	8.1240	4.0412
18	324	5832	4.2426	2.6207		67	4489	300763	8.1854	4.0615
19	361	6859	4.3589	2.6684		68	4624	314432	8.2462	4.0817
20	400	8000	4.4721	2.7144		69	4761	328509	8.3066	4.1016
21	441	9261	4.5826	2.7589		70	4900	343000	8.3666	4.1213
22	484	10648	4.6904	2.8020		71	5041	357911	8.4261	4.1408
23	529	12167	4.7958	2.8439		72	5184	373248	8.4853	4.1602
24	576	13824	4.8990	2.8845		73	5329	389017	8.5440	4.1793
25	625	15625	5.0000	2.9240		74	5476	405224	8.6023	4.1983
26	676	17576	5.0990	2.9625		75	5625	421875	8.6603	4.2172
27	729	19683	5.1962	3.0000		76	5776	438976	8.7178	4.2358
28	784	21952	5.2915	3.0366		77	5929	456533	8.7750	4.2543
29	841	24389	5.3852	3.0723		78	6084	474552	8.8318	4.2727
30	900	27000	5.4772	3.1072		79	6241	493039	8.8882	4.2908
31	961	29791	5.5678	3.1414		80	6400	512000	8.9443	4.3089
32	1024	32768	5.6569	3.1748		81	6561	531441	9.0000	4.3267
33	1089	35937	5.7446	3.2075		82	6724	551368	9.0554	4.3445
34	1156	39304	5.8310	3.2396		83	6889	571787	9.1104	4.3621
35	1225	42875	5.9161	3.2711		84	7056	592704	9.1652	4.3795
36	1296	46656	6.0000	3.3019		85	7225	614125	9.2195	4.3968
37	1369	50653	6.0828	3.3322		86	7396	636056	9.2736	4.4140
38	1444	54872	6.1644	3.3620		87	7569	658503	9.3274	4.4310
39	1521	59319	6.2450	3.3912		88	7744	681472	9.3808	4.4480
40	1600	64000	6.3246	3.4200		89	7921	704969	9.4340	4.4647
41	1681	68921	6.4031	3.4482		90	8100	729000	9.4868	4.4814
42	1764	74088	6.4807	3.4760		91	8281	753571	9.5394	4.4979
43	1849	79507	6.5574	3.5034		92	8464	778688	9.5917	4.5144
44	1936	85184	6.6332	3.5303		93	8649	804357	9.6437	4.5307
45	2025	91125	6.7082	3.5569		94	8836	830584	9.6954	4.5468
46	2116	97336	6.7823	3.5830		95	9025	857375	9.7468	4.5629
47	2209	103823	6.8557	3.6088		96	9216	884736	9.7980	4.5789
48	2304	110592	6.9282	3.6342		97	9409	912673	9.8489	4.5947
49	2401	117649	7.0000	3.6593		98	9604	941192	9.8995	4.6104
						99	9801	970299	9.9499	4.6261

No.	Square	Cube	Square Root	Cube Root		No.	Square	Cube	Square Root	Cube Root
100	10000	1000000	10.0000	4.6416		150	22500	3375000	12.2474	5.3133
101	10201	1030301	10.0499	4.6570		151	22801	3442951	12.2882	5.3251
102	10404	1061208	10.0995	4.6723		152	23104	3511808	12.3288	5.3368
103	10609	1092727	10.1489	4.6875		153	23409	3581577	12.3693	5.3485
104	10816	1124864	10.1980	4.7027		154	23716	3652264	12.4097	5.3601
105	11025	1157625	10.2470	4.7177		155	24025	3723875	12.4499	5.3717
106	11236	1191016	10.2956	4.7326		156	24336	3796416	12.4900	5.3832
107	11449	1225043	10.3441	4.7475		157	24649	3869893	12.5300	5.3947
108	11664	1259712	10.3923	4.7622		158	24964	3944312	12.5698	5.4061
109	11881	1295029	10.4403	4.7769		159	25281	4019679	12.6095	5.4175
110	12100	1331000	10.4881	4.7914		160	25600	4096000	12.6491	5.4288
111	12321	1367631	10.5357	4.8059		161	25921	4173281	12.6886	5.4401
112	12544	1404928	10.5830	4.8203		162	26244	4251528	12.7279	5.4514
113	12769	1442897	10.6301	4.8346		163	26569	4330747	12.7671	5.4626
114	12996	1481544	10.6771	4.8488		164	26896	4410944	12.8062	5.4737
115	13225	1520875	10.7238	4.8629		165	27225	4492125	12.8452	5.4848
116	13456	1560896	10.7703	4.8770		166	27556	4574296	12.8841	5.4959
117	13689	1601613	10.8167	4.8910		167	27889	4657463	12.9228	5.5069
118	13924	1643032	10.8628	4.9049		168	28224	4741632	12.9615	5.5178
119	14161	1685159	10.9087	4.9187		169	28561	4826809	13.0000	5.5288
120	14400	1728000	10.9545	4.9324		170	28900	4913000	13.0384	5.5397
121	14641	1771561	11.0000	4.9461		171	29241	5000211	13.0767	5.5505
122	14884	1815848	11.0454	4.9597		172	29584	5088448	13.1149	5.5613
123	15129	1860867	11.0905	4.9732		173	29929	5177717	13.1529	5.5721
124	15376	1906624	11.1355	4.9866		174	30276	5268024	13.1909	5.5828
125	15625	1953125	11.1803	5.0000		175	30625	5359375	13.2288	5.5934
126	15876	2000376	11.2250	5.0133		176	30976	5451776	13.2665	5.6041
127	16129	2048383	11.2694	5.0265		177	31329	5545233	13.3041	5.6147
128	16384	2097152	11.3137	5.0397		178	31684	5639752	13.3417	5.6252
129	16641	2146689	11.3578	5.0528		179	32041	5735339	13.3791	5.6357
130	16900	2197000	11.4018	5.0658		180	32400	5832000	13.4164	5.6462
131	17161	2248091	11.4455	5.0788		181	32761	5929741	13.4536	5.6567
132	17424	2299968	11.4891	5.0916		182	33124	6028568	13.4907	5.6671
133	17689	2352637	11.5326	5.1045		183	33489	6128487	13.5277	5.6774
134	17956	2406104	11.5758	5.1172		184	33856	6229504	13.5647	5.6877
135	18225	2460375	11.6190	5.1299		185	34225	6331625	13.6015	5.6980
136	18496	2515456	11.6619	5.1426		186	34596	6434856	13.6382	5.7083
137	18769	2571353	11.7047	5.1551		187	34969	6539203	13.6748	5.7185
138	19044	2628072	11.7473	5.1676		188	35344	6644672	13.7113	5.7287
139	19321	2685619	11.7898	5.1801		189	35721	6751269	13.7477	5.7388
140	19600	2744000	11.8322	5.1925		190	36100	6859000	13.7840	5.7489
141	19881	2803221	11.8743	5.2048		191	36481	6967871	13.8203	5.7590
142	20164	2863288	11.9164	5.2171		192	36864	7077888	13.8564	5.7690
143	20449	2924207	11.9583	5.2293		193	37249	7189057	13.8924	5.7790
144	20736	2985984	12.0000	5.2415		194	37636	7301384	13.9284	5.7890
145	21025	3048625	12.0416	5.2536		195	38025	7414875	13.9642	5.7989
146	21316	3112136	12.0830	5.2656		196	38416	7529536	14.0000	5.8088
147	21609	3176523	12.1244	5.2776		197	38809	7645373	14.0357	5.8186
148	21904	3241792	12.1655	5.2896		198	39204	7762392	14.0712	5.8285
149	22201	3307949	12.2066	5.3015		199	39601	7880599	14.1067	5.8383

No.	Square	Cube	Square Root	Cube Root
200	40000	8000000	14.1421	5.8480
201	40401	8120601	14.1774	5.8578
202	40804	8242408	14.2127	5.8675
203	41209	8365427	14.2478	5.8771
204	41616	8489664	14.2829	5.8868
205	42025	8615125	14.3178	5.8964
206	42436	8741816	14.3527	5.9059
207	42849	8869743	14.3875	5.9155
208	43264	8998912	14.4222	5.9250
209	43681	9129329	14.4568	5.9345
210	44100	9261000	14.4914	5.9439
211	44521	9393931	14.5258	5.9533
212	44944	9528128	14.5602	5.9627
213	45369	9663597	14.5945	5.9721
214	45796	9800344	14.6287	5.9814
215	46225	9938375	14.6629	5.9907
216	46656	10077696	14.6969	6.0000
217	47089	10218313	14.7309	6.0092
218	47524	10360232	14.7648	6.0185
219	47961	10503459	14.7986	6.0277
220	48400	10648000	14.8324	6.0368
221	48841	10793861	14.8661	6.0459
222	49284	10941048	14.8997	6.0550
223	49729	11089567	14.9332	6.0641
224	50176	11239424	14.9666	6.0732
225	50625	11390625	15.0000	6.0822
226	51076	11543176	15.0333	6.0912
227	51529	11697083	15.0665	6.1002
228	51984	11852352	15.0997	6.1091
229	52441	12008989	15.1327	6.1180
230	52900	12167000	15.1658	6.1269
231	53361	12326391	15.1987	6.1358
232	53824	12487168	15.2315	6.1446
233	54289	12649337	15.2643	6.1534
234	54756	12812904	15.2971	6.1622
235	55225	12977875	15.3297	6.1710
236	55696	13144256	15.3623	6.1797
237	56169	13312053	15.3948	6.1885
238	56644	13481272	15.4272	6.1972
239	57121	13651919	15.4596	6.2058
240	57600	13824000	15.4919	6.2145
241	58081	13997521	15.5242	6.2231
242	58564	14172488	15.5563	6.2317
243	59049	14348907	15.5885	6.2403
244	59536	14526784	15.6205	6.2488
245	60025	14706125	15.6525	6.2573
246	60516	14886936	15.6844	6.2658
247	61009	15069223	15.7162	6.2743
248	61504	15252992	15.7480	6.2828
249	62001	15438249	15.7797	6.2912

No.	Square	Cube	Square Root	Cube Root
250	62500	15625000	15.8114	6.2996
251	63001	15813251	15.8430	6.3080
252	63504	16003008	15.8745	6.3164
253	64009	16194277	15.9060	6.3247
254	64516	16387064	15.9374	6.3330
255	65025	16581375	15.9687	6.3413
256	65536	16777216	16.0000	6.3496
257	66049	16974593	16.0312	6.3579
258	66564	17173512	16.0624	6.3661
259	67081	17373979	16.0935	6.3743
260	67600	17576000	16.1245	6.3825
261	68121	17779581	16.1555	6.3907
262	68644	17984728	16.1864	6.3988
263	69169	18191447	16.2173	6.4070
264	69696	18399744	16.2481	6.4151
265	70225	18609625	16.2788	6.4232
266	70756	18821096	16.3095	6.4312
267	71289	19034163	16.3401	6.4393
268	71824	19248832	16.3707	6.4473
269	72361	19465109	16.4012	6.4553
270	72900	19683000	16.4317	6.4633
271	73441	19902511	16.4621	6.4713
272	73984	20123648	16.4924	6.4792
273	74529	20346417	16.5227	6.4872
274	75076	20570824	16.5529	6.4951
275	75625	20796875	16.5831	6.5030
276	76176	21024576	16.6132	6.5108
277	76729	21253933	16.6433	6.5187
278	77284	21484952	16.6733	6.5265
279	77841	21717639	16.7033	6.5343
280	78400	21952000	16.7332	6.5421
281	78961	22188041	16.7631	6.5499
282	79524	22425768	16.7929	6.5577
283	80089	22665187	16.8226	6.5654
284	80656	22906304	16.8523	6.5731
285	81225	23149125	16.8819	6.5808
286	81796	23393656	16.9115	6.5885
287	82369	23639903	16.9411	6.5962
288	82944	23887872	16.9706	6.6039
289	83521	24137569	17.0000	6.6115
290	84100	24389000	17.0294	6.6191
291	84681	24642171	17.0587	6.6267
292	85264	24897088	17.0880	6.6343
293	85849	25153757	17.1172	6.6419
294	86436	25412184	17.1464	6.6494
295	87025	25672375	17.1756	6.6569
296	87616	25934336	17.2047	6.6644
297	88209	26198073	17.2337	6.6719
298	88804	26463592	17.2627	6.6794
299	89401	26730899	17.2916	6.6869

No.	Square	Cube	Square Root	Cube Root		No.	Square	Cube	Square Root	Cube Root
300	90000	27000000	17.3205	6.6943		350	122500	42875000	18.7083	7.0473
301	90601	27270901	17.3494	6.7018		351	123201	43243551	18.7350	7.0540
302	91204	27543608	17.3781	6.7092		352	123904	43614208	18.7617	7.0607
303	91809	27818127	17.4069	6.7166		353	124609	43986977	18.7883	7.0674
304	92416	28094464	17.4356	6.7240		354	125316	44361864	18.8149	7.0740
305	93025	28372625	17.4642	6.7313		355	126025	44738875	18.8414	7.0807
306	93636	28652616	17.4929	6.7387		356	126736	45118016	18.8680	7.0873
307	94249	28934443	17.5214	6.7460		357	127449	45499293	18.8944	7.0940
308	94864	29218112	17.5499	6.7533		358	128164	45882712	18.9209	7.1006
309	95481	29503629	17.5784	6.7606		359	128881	46268279	18.9473	7.1072
310	96100	29791000	17.6068	6.7679		360	129600	46656000	18.9737	7.1138
311	96721	30080231	17.6352	6.7752		361	130321	47045881	19.0000	7.1204
312	97344	30371328	17.6635	6.7824		362	131044	47437928	19.0263	7.1269
313	97969	30664297	17.6918	6.7897		363	131769	47832147	19.0526	7.1335
314	98596	30959144	17.7200	6.7969		364	132496	48228544	19.0788	7.1400
315	99225	31255875	17.7482	6.8041		365	133225	48627125	19.1050	7.1466
316	99856	31554496	17.7764	6.8113		366	133956	49027896	19.1311	7.1531
317	100489	31855013	17.8045	6.8185		367	134689	49430863	19.1572	7.1596
318	101124	32157432	17.8326	6.8256		368	135424	49836032	19.1833	7.1661
319	101761	32461759	17.8606	6.8328		369	136161	50243409	19.2094	7.1726
320	102400	32768000	17.8885	6.8399		370	136900	50653000	19.2354	7.1791
321	103041	33076161	17.9165	6.8470		371	137641	51064811	19.2614	7.1855
322	103684	33386248	17.9444	6.8541		372	138384	51478848	19.2873	7.1920
323	104329	33698267	17.9722	6.8612		373	139129	51895117	19.3132	7.1984
324	104976	34012224	18.0000	6.8683		374	139876	52313624	19.3391	7.2048
325	105625	34328125	18.0278	6.8753		375	140625	52734375	19.3649	7.2112
326	106276	34645976	18.0555	6.8824		376	141376	53157376	19.3907	7.2177
327	106929	34965783	18.0831	6.8894		377	142129	53582633	19.4165	7.2240
328	107584	35287552	18.1108	6.8964		378	142884	54010152	19.4422	7.2304
329	108241	35611289	18.1384	6.9034		379	143641	54439939	19.4679	7.2368
330	108900	35937000	18.1659	6.9104		380	144400	54872000	19.4936	7.2432
331	109561	36264691	18.1934	6.9174		381	145161	55306341	19.5192	7.2495
332	110224	36594368	18.2209	6.9244		382	145924	55742968	19.5448	7.2558
333	110889	36926037	18.2483	6.9313		383	146689	56181887	19.5704	7.2622
334	111556	37259704	18.2757	6.9382		384	147456	56623104	19.5959	7.2685
335	112225	37595375	18.3030	6.9451		385	148225	57066625	19.6214	7.2748
336	112896	37933056	18.3303	6.9521		386	148996	57512456	19.6469	7.2811
337	113569	38272753	18.3576	6.9589		387	149769	57960603	19.6723	7.2874
338	114244	38614472	18.3848	6.9658		388	150544	58411072	19.6977	7.2936
339	114921	38958219	18.4120	6.9727		389	151321	58863869	19.7231	7.2999
340	115600	39304000	18.4391	6.9795		390	152100	59319000	19.7484	7.3061
341	116281	39651821	18.4662	6.9864		391	152881	59776471	19.7737	7.3124
342	116964	40001688	18.4932	6.9932		392	153664	60236288	19.7990	7.3186
343	117649	40353607	18.5203	7.0000		393	154449	60698457	19.8242	7.3248
344	118336	40707584	18.5472	7.0068		394	155236	61162984	19.8494	7.3310
345	119025	41063625	18.5742	7.0136		395	156025	61629875	19.8746	7.3372
346	119716	41421736	18.6011	7.0203		396	156816	62099136	19.8997	7.3434
347	120409	41781923	18.6279	7.0271		397	157609	62570773	19.9249	7.3496
348	121104	42144192	18.6548	7.0338		398	158404	63044792	19.9499	7.3558
349	121801	42508549	18.6815	7.0406		399	159201	63521199	19.9750	7.3619

No.	Square	Cube	Square Root	Cube Root	No.	Square	Cube	Square Root	Cube Root
400	160000	64000000	20.0000	7.3681	450	202500	91125000	21.2132	7.6631
401	160801	64481201	20.0250	7.3742	451	203401	91733851	21.2368	7.6688
402	161604	64964808	20.0499	7.3803	452	204304	92345408	21.2603	7.6744
403	162409	65450827	20.0749	7.3864	453	205209	92959677	21.2838	7.6801
404	163216	65939264	20.0998	7.3925	454	206116	93576664	21.3073	7.6857
405	164025	66430125	20.1246	7.3986	455	207025	94196375	21.3307	7.6914
406	164836	66923416	20.1494	7.4047	456	207936	94818816	21.3542	7.6970
407	165649	67419143	20.1742	7.4108	457	208849	95443993	21.3776	7.7026
408	166464	67917312	20.1990	7.4169	458	209764	96071912	21.4009	7.7082
409	167281	68417929	20.2237	7.4229	459	210681	96702579	21.4243	7.7138
410	168100	68921000	20.2485	7.4290	460	211600	97336000	21.4476	7.7194
411	168921	69426531	20.2731	7.4350	461	212521	97972181	21.4709	7.7250
412	169744	69934528	20.2978	7.4410	462	213444	98611128	21.4942	7.7306
413	170569	70444997	20.3224	7.4470	463	214369	99252847	21.5174	7.7362
414	171396	70957944	20.3470	7.4530	464	215296	99897344	21.5407	7.7418
415	172225	71473375	20.3715	7.4590	465	216225	100544625	21.5639	7.7473
416	173056	71991296	20.3961	7.4650	466	217156	101194696	21.5870	7.7529
417	173889	72511713	20.4206	7.4710	467	218089	101847563	21.6102	7.7584
418	174724	73034632	20.4450	7.4770	468	219024	102503232	21.6333	7.7639
419	175561	73560059	20.4695	7.4829	469	219961	103161709	21.6564	7.7695
420	176400	74088000	20.4939	7.4889	470	220900	103823000	21.6795	7.7750
421	177241	74618461	20.5183	7.4948	471	221841	104487111	21.7025	7.7805
422	178084	75151448	20.5426	7.5007	472	222784	105154048	21.7256	7.7860
423	178929	75686967	20.5670	7.5067	473	223729	105823817	21.7486	7.7915
424	179776	76225024	20.5913	7.5126	474	224676	106496424	21.7715	7.7970
425	180625	76765625	20.6155	7.5185	475	225625	107171875	21.7945	7.8025
426	181476	77308776	20.6398	7.5244	476	226576	107850176	21.8174	7.8079
427	182329	77854483	20.6640	7.5302	477	227529	108531333	21.8403	7.8134
428	183184	78402752	20.6882	7.5361	478	228484	109215352	21.8632	7.8188
429	184041	78953589	20.7123	7.5420	479	229441	109902239	21.8861	7.8243
430	184900	79507000	20.7364	7.5478	480	230400	110592000	21.9089	7.8297
431	185761	80062991	20.7605	7.5537	481	231361	111284641	21.9317	7.8352
432	186624	80621568	20.7846	7.5595	482	232324	111980168	21.9545	7.8406
433	187489	81182737	20.8087	7.5654	483	233289	112678587	21.9773	7.8460
434	188356	81746504	20.8327	7.5712	484	234256	113379904	22.0000	7.8514
435	189225	82312875	20.8567	7.5770	485	235225	114084125	22.0227	7.8568
436	190096	82881856	20.8806	7.5828	486	236196	114791256	22.0454	7.8622
437	190969	83453453	20.9045	7.5886	487	237169	115501303	22.0681	7.8676
438	191844	84027672	20.9284	7.5944	488	238144	116214272	22.0907	7.8730
439	192721	84604519	20.9523	7.6001	489	239121	116930169	22.1133	7.8784
440	193600	85184000	20.9762	7.6059	490	240100	117649000	22.1359	7.8837
441	194481	85766121	21.0000	7.6117	491	241081	118370771	22.1585	7.8891
442	195364	86350888	21.0238	7.6174	492	242064	119095488	22.1811	7.8944
443	196249	86938307	21.0476	7.6232	493	243049	119823157	22.2036	7.8998
444	197136	87528384	21.0713	7.6289	494	244036	120553784	22.2261	7.9051
445	198025	88121125	21.0950	7.6346	495	245025	121287375	22.2486	7.9105
446	198916	88716536	21.1187	7.6403	496	246016	122023936	22.2711	7.9158
447	199809	89314623	21.1424	7.6460	497	247009	122763473	22.2935	7.9211
448	200704	89915392	21.1660	7.6517	498	248004	123505992	22.3159	7.9264
449	201601	90518849	21.1896	7.6574	499	249001	124251499	22.3383	7.9317

No.	Square	Cube	Square Root	Cube Root	No.	Square	Cube	Square Root	Cube Root
500	250000	125000000	22.3607	7.9370	550	302500	166375000	23.4521	8.1932
501	251001	125751501	22.3830	7.9423	551	303601	167284151	23.4734	8.1982
502	252004	126506008	22.4054	7.9476	552	304704	168196608	23.4947	8.2031
503	253009	127263527	22.4277	7.9528	553	305809	169112377	23.5160	8.2081
504	254016	128024064	22.4499	7.9581	554	306916	170031464	23.5372	8.2130
505	255025	128787625	22.4722	7.9634	555	308025	170953875	23.5584	8.2180
506	256036	129554216	22.4944	7.9686	556	309136	171879616	23.5797	8.2229
507	257049	130323843	22.5167	7.9739	557	310249	172808693	23.6008	8.2278
508	258064	131096512	22.5389	7.9791	558	311364	173741112	23.6220	8.2327
509	259081	131872229	22.5610	7.9843	559	312481	174676879	23.6432	8.2377
510	260100	132651000	22.5832	7.9896	560	313600	175616000	23.6643	8.2426
511	261121	133432831	22.6053	7.9948	561	314721	176558481	23.6854	8.2475
512	262144	134217728	22.6274	8.0000	562	315844	177504328	23.7065	8.2524
513	263169	135005697	22.6495	8.0052	563	316969	178453547	23.7276	8.2573
514	264196	135796744	22.6716	8.0104	564	318096	179406144	23.7487	8.2621
515	265225	136590875	22.6936	8.0156	565	319225	180362125	23.7697	8.2670
516	266256	137388096	22.7156	8.0208	566	320356	181321496	23.7908	8.2719
517	267289	138188413	22.7376	8.0260	567	321489	182284263	23.8118	8.2768
518	268324	138991832	22.7596	8.0311	568	322624	183250432	23.8328	8.2816
519	269361	139798359	22.7816	8.0363	569	323761	184220009	23.8537	8.2865
520	270400	140608000	22.8035	8.0415	570	324900	185193000	23.8747	8.2913
521	271441	141420761	22.8254	8.0466	571	326041	186169411	23.8956	8.2962
522	272484	142236648	22.8473	8.0517	572	327184	187149248	23.9165	8.3010
523	273529	143055667	22.8692	8.0569	573	328329	188132517	23.9374	8.3059
524	274576	143877824	22.8910	8.0620	574	329476	189119224	23.9583	8.3107
525	275625	144703125	22.9129	8.0671	575	330625	190109375	23.9792	8.3155
526	276676	145531576	22.9347	8.0723	576	331776	191102976	24.0000	8.3203
527	277729	146363183	22.9565	8.0774	577	332929	192100033	24.0208	8.3251
528	278784	147197952	22.9783	8.0825	578	334084	193100552	24.0416	8.3300
529	279841	148035889	23.0000	8.0876	579	335241	194104539	24.0624	8.3348
530	280900	148877000	23.0217	8.0927	580	336400	195112000	24.0832	8.3396
531	281961	149721291	23.0434	8.0978	581	337561	196122941	24.1039	8.3443
532	283024	150568768	23.0651	8.1028	582	338724	197137368	24.1247	8.3491
533	284089	151419437	23.0868	8.1079	583	339889	198155287	24.1454	8.3539
534	285156	152273304	23.1084	8.1130	584	341056	199176704	24.1661	8.3587
535	286225	153130375	23.1301	8.1180	585	342225	200201625	24.1868	8.3634
536	287296	153990656	23.1517	8.1231	586	343396	201230056	24.2074	8.3682
537	288369	154854153	23.1733	8.1281	587	344569	202262003	24.2281	8.3730
538	289444	155720872	23.1948	8.1332	588	345744	203297472	24.2487	8.3777
539	290521	156590819	23.2164	8.1382	589	346921	204336469	24.2693	8.3825
540	291600	157464000	23.2379	8.1433	590	348100	205379000	24.2899	8.3872
541	292681	158340421	23.2594	8.1483	591	349281	206425071	24.3105	8.3919
542	293764	159220088	23.2809	8.1533	592	350464	207474688	24.3311	8.3967
543	294849	160103007	23.3024	8.1583	593	351649	208527857	24.3516	8.4014
544	295936	160989184	23.3238	8.1633	594	352836	209584584	24.3721	8.4061
545	297025	161878625	23.3452	8.1683	595	354025	210644875	24.3926	8.4108
546	298116	162771336	23.3666	8.1733	596	355216	211708736	24.4131	8.4155
547	299209	163667323	23.3880	8.1783	597	356409	212776173	24.4336	8.4202
548	300304	164566592	23.4094	8.1833	598	357604	213847192	24.4540	8.4249
549	301401	165469149	23.4307	8.1882	599	358801	214921799	24.4745	8.4296

No.	Square	Cube	Square Root	Cube Root		No.	Square	Cube	Square Root	Cube Root
600	360000	216000000	24.4949	8.4343		650	422500	274625000	25.4951	8.6624
601	361201	217081801	24.5153	8.4390		651	423801	275894451	25.5147	8.6668
602	362404	218167208	24.5357	8.4437		652	425104	277167808	25.5343	8.6713
603	363609	219256227	24.5561	8.4484		653	426409	278445077	25.5539	8.6757
604	364816	220348864	24.5764	8.4530		654	427716	279726264	25.5734	8.6801
605	366025	221445125	24.5967	8.4577		655	429025	281011375	25.5930	8.6845
606	367236	222545016	24.6171	8.4623		656	430336	282300416	25.6125	8.6890
607	368449	223648543	24.6374	8.4670		657	431649	283593393	25.6320	8.6934
608	369664	224755712	24.6577	8.4716		658	432964	284890312	25.6515	8.6978
609	370881	225866529	24.6779	8.4763		659	434281	286191179	25.6710	8.7022
610	372100	226981000	24.6982	8.4809		660	435600	287496000	25.6905	8.7066
611	373321	228099131	24.7184	8.4856		661	436921	288804781	25.7099	8.7110
612	374544	229220928	24.7386	8.4902		662	438244	290117528	25.7294	8.7154
613	375769	230346397	24.7588	8.4948		663	439569	291434247	25.7488	8.7198
614	376996	231475544	24.7790	8.4994		664	440896	292754944	25.7682	8.7241
615	378225	232608375	24.7992	8.5040		665	442225	294079625	25.7876	8.7285
616	379456	233744896	24.8193	8.5086		666	443556	295408296	25.8070	8.7329
617	380689	234885113	24.8395	8.5132		667	444889	296740963	25.8263	8.7373
618	381924	236029032	24.8596	8.5178		668	446224	298077632	25.8457	8.7416
619	383161	237176659	24.8797	8.5224		669	447561	299418309	25.8650	8.7460
620	384400	238328000	24.8998	8.5270		670	448900	300763000	25.8844	8.7503
621	385641	239483061	24.9199	8.5316		671	450241	302111711	25.9037	8.7547
622	386884	240641848	24.9399	8.5362		672	451584	303464448	25.9230	8.7590
623	388129	241804367	24.9600	8.5408		673	452929	304821217	25.9422	8.7634
624	389376	242970624	24.9800	8.5453		674	454276	306182024	25.9615	8.7677
625	390625	244140625	25.0000	8.5499		675	455625	307546875	25.9808	8.7721
626	391876	245314376	25.0200	8.5544		676	456976	308915776	26.0000	8.7764
627	393129	246491883	25.0400	8.5590		677	458329	310288733	26.0192	8.7807
628	394384	247673152	25.0599	8.5635		678	459684	311665752	26.0384	8.7850
629	395641	248858189	25.0799	8.5681		679	461041	313046839	26.0576	8.7893
630	396900	250047000	25.0998	8.5726		680	462400	314432000	26.0768	8.7937
631	398161	251239591	25.1197	8.5772		681	463761	315821241	26.0960	8.7980
632	399424	252435968	25.1396	8.5817		682	465124	317214568	26.1151	8.8023
633	400689	253636137	25.1595	8.5862		683	466489	318611987	26.1343	8.8066
634	401956	254840104	25.1794	8.5907		684	467856	320013504	26.1534	8.8109
635	403225	256047875	25.1992	8.5952		685	469225	321419125	26.1725	8.8152
636	404496	257259456	25.2190	8.5997		686	470596	322828856	26.1916	8.8194
637	405769	258474853	25.2389	8.6043		687	471969	324242703	26.2107	8.8237
638	407044	259694072	25.2587	8.6088		688	473344	325660672	26.2298	8.8280
639	408321	260917119	25.2784	8.6132		689	474721	327082769	26.2488	8.8323
640	409600	262144000	25.2982	8.6177		690	476100	328509000	26.2679	8.8366
641	410881	263374721	25.3180	8.6222		691	477481	329939371	26.2869	8.8408
642	412164	264609288	25.3377	8.6267		692	478864	331373888	26.3059	8.8451
643	413449	265847707	25.3574	8.6312		693	480249	332812557	26.3249	8.8493
644	414736	267089984	25.3772	8.6357		694	481636	334255384	26.3439	8.8536
645	416025	268336125	25.3969	8.6401		695	483025	335702375	26.3629	8.8578
646	417316	269586136	25.4165	8.6446		696	484416	337153536	26.3818	8.8621
647	418609	270840023	25.4362	8.6490		697	485809	338608873	26.4008	8.8663
648	419904	272097792	25.4558	8.6535		698	487204	340068392	26.4197	8.8706
649	421201	273359449	25.4755	8.6579		699	488601	341532099	26.4386	8.8748

No.	Square	Cube	Square Root	Cube Root	No.	Square	Cube	Square Root	Cube Root
700	490000	343000000	26.4575	8.8790	750	562500	421875000	27.3861	9.0856
701	491401	344472101	26.4764	8.8833	751	564001	423564751	27.4044	9.0896
702	492804	345948408	26.4953	8.8875	752	565504	425259008	27.4226	9.0937
703	494209	347428927	26.5141	8.8917	753	567009	426957777	27.4408	9.0977
704	495616	348913664	26.5330	8.8959	754	568516	428661064	27.4591	9.1017
705	497025	350402625	26.5518	8.9001	755	570025	430368875	27.4773	9.1057
706	498436	351895816	26.5707	8.9043	756	571536	432081216	27.4955	9.1098
707	499849	353393243	26.5895	8.9085	757	573049	433798093	27.5136	9.1138
708	501264	354894912	26.6083	8.9127	758	574564	435519512	27.5318	9.1178
709	502681	356400829	26.6271	8.9169	759	576081	437245479	27.5500	9.1218
710	504100	357911000	26.6458	8.9211	760	577600	438976000	27.5681	9.1258
711	505521	359425431	26.6646	8.9253	761	579121	440711081	27.5862	9.1298
712	506944	360944128	26.6833	8.9295	762	580644	442450728	27.6043	9.1338
713	508369	362467097	26.7021	8.9337	763	582169	444194947	27.6225	9.1378
714	509796	363994344	26.7208	8.9378	764	583696	445943744	27.6405	9.1418
715	511225	365525875	26.7395	8.9420	765	585225	447697125	27.6586	9.1458
716	512656	367061696	26.7582	8.9462	766	586756	449455096	27.6767	9.1498
717	514089	368601813	26.7769	8.9503	767	588289	451217663	27.6948	9.1537
718	515524	370146232	26.7955	8.9545	768	589824	452984832	27.7128	9.1577
719	516961	371694959	26.8142	8.9587	769	591361	454756609	27.7308	9.1617
720	518400	373248000	26.8328	8.9628	770	592900	456533000	27.7489	9.1657
721	519841	374805361	26.8514	8.9670	771	594441	458314011	27.7669	9.1696
722	521284	376367048	26.8701	8.9711	772	595984	460099648	27.7849	9.1736
723	522729	377933067	26.8887	8.9752	773	597529	461889917	27.8029	9.1775
724	524176	379503424	26.9072	8.9794	774	599076	463684824	27.8209	9.1815
725	525625	381078125	26.9258	8.9835	775	600625	465484375	27.8388	9.1855
726	527076	382657176	26.9444	8.9876	776	602176	467288576	27.8568	9.1894
727	528529	384240583	26.9629	8.9918	777	603729	469097433	27.8747	9.1933
728	529984	385828352	26.9815	8.9959	778	605284	470910952	27.8927	9.1973
729	531441	387420489	27.0000	9.0000	779	606841	472729139	27.9106	9.2012
730	532900	389017000	27.0185	9.0041	780	608400	474552000	27.9285	9.2052
731	534361	390617891	27.0370	9.0082	781	609961	476379541	27.9464	9.2091
732	535824	392223168	27.0555	9.0123	782	611524	478211768	27.9643	9.2130
733	537289	393832837	27.0740	9.0164	783	613089	480048687	27.9821	9.2170
734	538756	395446904	27.0924	9.0205	784	614656	481890304	28.0000	9.2209
735	540225	397065375	27.1109	9.0246	785	616225	483736625	28.0179	9.2248
736	541696	398688256	27.1293	9.0287	786	617796	485587656	28.0357	9.2287
737	543169	400315553	27.1477	9.0328	787	619369	487443403	28.0535	9.2326
738	544644	401947272	27.1662	9.0369	788	620944	489303872	28.0713	9.2365
739	546121	403583419	27.1846	9.0410	789	622521	491169069	28.0891	9.2404
740	547600	405224000	27.2029	9.0450	790	624100	493039000	28.1069	9.2443
741	549081	406869021	27.2213	9.0491	791	625681	494913671	28.1247	9.2482
742	550564	408518488	27.2397	9.0532	792	627264	496793088	28.1425	9.2521
743	552049	410172407	27.2580	9.0572	793	628849	498677257	28.1603	9.2560
744	553536	411830784	27.2764	9.0613	794	630436	500566184	28.1780	9.2599
745	555025	413493625	27.2947	9.0654	795	632025	502459875	28.1957	9.2638
746	556516	415160936	27.3130	9.0694	796	633616	504358336	28.2135	9.2677
747	558009	416832723	27.3313	9.0735	797	635209	506261573	28.2312	9.2716
748	559504	418508992	27.3496	9.0775	798	636804	508169592	28.2489	9.2754
749	561001	420189749	27.3679	9.0816	799	638401	510082399	28.2666	9.2793

No.	Square	Cube	Square Root	Cube Root	No.	Square	Cube	Square Root	Cube Root
800	640000	512000000	28.2843	9.2832	850	722500	614125000	29.1548	9.4727
801	641601	513922401	28.3019	9.2870	851	724201	616295051	29.1719	9.4764
802	643204	515849608	28.3196	9.2909	852	725904	618470208	29.1890	9.4801
803	644809	517781627	28.3373	9.2948	853	727609	620650477	29.2062	9.4838
804	646416	519718464	28.3549	9.2986	854	729316	622835864	29.2233	9.4875
805	648025	521660125	28.3725	9.3025	855	731025	625026375	29.2404	9.4912
806	649636	523606616	28.3901	9.3063	856	732736	627222016	29.2575	9.4949
807	651249	525557943	28.4077	9.3102	857	734449	629422793	29.2746	9.4986
808	652864	527514112	28.4253	9.3140	858	736164	631628712	29.2916	9.5023
809	654481	529475129	28.4429	9.3179	859	737881	633839779	29.3087	9.5060
810	656100	531441000	28.4605	9.3217	860	739600	636056000	29.3258	9.5097
811	657721	533411731	28.4781	9.3255	861	741321	638277381	29.3428	9.5134
812	659344	535387328	28.4956	9.3294	862	743044	640503928	29.3598	9.5171
813	660969	537367797	28.5132	9.3332	863	744769	642735647	29.3769	9.5207
814	662596	539353144	28.5307	9.3370	864	746496	644972544	29.3939	9.5244
815	664225	541343375	28.5482	9.3408	865	748225	647214625	29.4109	9.5281
816	665856	543338496	28.5657	9.3447	866	749956	649461896	29.4279	9.5317
817	667489	545338513	28.5832	9.3485	867	751689	651714363	29.4449	9.5354
818	669124	547343432	28.6007	9.3523	868	753424	653972032	29.4618	9.5391
819	670761	549353259	28.6182	9.3561	869	755161	656234909	29.4788	9.5427
820	672400	551368000	28.6356	9.3599	870	756900	658503000	29.4958	9.5464
821	674041	553387661	28.6531	9.3637	871	758641	660776311	29.5127	9.5501
822	675684	555412248	28.6705	9.3675	872	760384	663054848	29.5296	9.5537
823	677329	557441767	28.6880	9.3713	873	762129	665338617	29.5466	9.5574
824	678976	559476224	28.7054	9.3751	874	763876	667627624	29.5635	9.5610
825	680625	561515625	28.7228	9.3789	875	765625	669921875	29.5804	9.5647
826	682276	563559976	28.7402	9.3827	876	767376	672221376	29.5973	9.5683
827	683929	565609283	28.7576	9.3865	877	769129	674526133	29.6142	9.5719
828	685584	567663552	28.7750	9.3902	878	770884	676836152	29.6311	9.5756
829	687241	569722789	28.7924	9.3940	879	772641	679151439	29.6479	9.5792
830	688900	571787000	28.8097	9.3978	880	774400	681472000	29.6648	9.5828
831	690561	573856191	28.8271	9.4016	881	776161	683797841	29.6816	9.5865
832	692224	575930368	28.8444	9.4053	882	777924	686128968	29.6985	9.5901
833	693889	578009537	28.8617	9.4091	883	779689	688465387	29.7153	9.5937
834	695556	580093704	28.8791	9.4129	884	781456	690807104	29.7321	9.5973
835	697225	582182875	28.8964	9.4166	885	783225	693154125	29.7489	9.6010
836	698896	584277056	28.9137	9.4204	886	784996	695506456	29.7658	9.6046
837	700569	586376253	28.9310	9.4241	887	786769	697864103	29.7825	9.6082
838	702244	588480472	28.9482	9.4279	888	788544	700227072	29.7993	9.6118
839	703921	590589719	28.9655	9.4316	889	790321	702595369	29.8161	9.6154
840	705600	592704000	28.9828	9.4354	890	792100	704969000	29.8329	9.6190
841	707281	594823321	29.0000	9.4391	891	793881	707347971	29.8496	9.6226
842	708964	596947688	29.0172	9.4429	892	795664	709732288	29.8664	9.6262
843	710649	599077107	29.0345	9.4466	893	797449	712121957	29.8831	9.6298
844	712336	601211584	29.0517	9.4503	894	799236	714516984	29.8998	9.6334
845	714025	603351125	29.0689	9.4541	895	801025	716917375	29.9166	9.6370
846	715716	605495736	29.0861	9.4578	896	802816	719323136	29.9333	9.6406
847	717409	607645423	29.1033	9.4615	897	804609	721734273	29.9500	9.6442
848	719104	609800192	29.1204	9.4652	898	806404	724150792	29.9666	9.6477
849	720801	611960049	29.1376	9.4690	899	808201	726572699	29.9833	9.6513

No.	Square	Cube	Square Root	Cube Root	No.	Square	Cube	Square Root	Cube Root
900	810000	729000000	30.0000	9.6549	950	902500	857375000	30.8221	9.8305
901	811801	731432701	30.0167	9.6585	951	904401	860035351	30.8383	9.8339
902	813604	733870808	30.0333	9.6620	952	906304	862801408	30.8545	9.8374
903	815409	736314327	30.0500	9.6656	953	908209	865523177	30.8707	9.8408
904	817216	738763264	30.0666	9.6692	954	910116	868250664	30.8869	9.8443
905	819025	741217625	30.0832	9.6727	955	912025	870983875	30.9031	9.8477
906	820836	743677416	30.0998	9.6763	956	913936	873722816	30.9192	9.8511
907	822649	746142643	30.1164	9.6799	957	915849	876467493	30.9354	9.8546
908	824464	748613312	30.1330	9.6834	958	917764	879217912	30.9516	9.8580
909	826281	751089429	30.1496	9.6870	959	919681	881974079	30.9677	9.8614
910	828100	753571000	30.1662	9.6905	960	921600	884736000	30.9839	9.8648
911	829921	756058031	30.1828	9.6941	961	923521	887503681	31.0000	9.8683
912	831744	758550528	30.1993	9.6976	962	925444	890277128	31.0161	9.8717
913	833569	761048497	30.2159	9.7012	963	927369	893056347	31.0322	9.8751
914	835396	763551944	30.2324	9.7047	964	929296	895841344	31.0483	9.8785
915	837225	766060875	30.2490	9.7082	965	931225	898632125	31.0644	9.8819
916	839056	768575296	30.2655	9.7118	966	933156	901428696	31.0805	9.8854
917	840889	771095213	30.2820	9.7153	967	935089	904231063	31.0966	9.8888
918	842724	773620632	30.2985	9.7188	968	937024	907039232	31.1127	9.8922
919	844561	776151559	30.3150	9.7224	969	938961	909853209	31.1288	9.8956
920	846400	778688000	30.3315	9.7259	970	940900	912673000	31.1448	9.8990
921	848241	781229961	30.3480	9.7294	971	942841	915498611	31.1609	9.9024
922	850084	783777448	30.3645	9.7329	972	944784	918330048	31.1769	9.9058
923	851929	786330467	30.3809	9.7364	973	946729	921167317	31.1929	9.9092
924	853776	788889024	30.3974	9.7400	974	948676	924010424	31.2090	9.9126
925	855625	791453125	30.4138	9.7435	975	950625	926859375	31.2250	9.9160
926	857476	794022776	30.4302	9.7470	976	952576	929714176	31.2410	9.9194
927	859329	796597983	30.4467	9.7505	977	954529	932574833	31.2570	9.9227
928	861184	799178752	30.4631	9.7540	978	956484	935441352	31.2730	9.9261
929	863041	801765089	30.4795	9.7575	979	958441	938313739	31.2890	9.9295
930	864900	804357000	30.4959	9.7610	980	960400	941192000	31.3050	9.9329
931	866761	806954491	30.5123	9.7645	981	962361	944076141	31.3209	9.9363
932	868624	809557568	30.5287	9.7680	982	964324	946966168	31.3369	9.9396
933	870489	812166237	30.5450	9.7715	983	966289	949862087	31.3528	9.9430
934	872356	814780504	30.5614	9.7750	984	968256	952763904	31.3688	9.9464
935	874225	817400375	30.5778	9.7785	985	970225	955671625	31.3847	9.9497
936	876096	820025856	30.5941	9.7819	986	972196	958585256	31.4006	9.9531
937	877969	822656953	30.6105	9.7854	987	974169	961504803	31.4166	9.9565
938	879844	825293672	30.6268	9.7889	988	976144	964430272	31.4325	9.9598
939	881721	827936019	30.6431	9.7924	989	978121	967361669	31.4484	9.9632
940	883600	830584000	30.6594	9.7959	990	980100	970299000	31.4643	9.9666
941	885481	833237621	30.6757	9.7993	991	982081	973242271	31.4802	9.9699
942	887364	835896888	30.6920	9.8028	992	984064	976191488	31.4960	9.9733
943	889249	838561807	30.7083	9.8063	993	986049	979146657	31.5119	9.9766
944	891136	841232384	30.7246	9.8097	994	988036	982107784	31.5278	9.9800
945	893025	843908625	30.7409	9.8132	995	990025	985074875	31.5436	9.9833
946	894916	846590536	30.7571	9.8167	996	992016	988047936	31.5595	9.9866
947	896809	849278123	30.7734	9.8201	997	994009	991026973	31.5753	9.9900
948	898704	851971392	30.7896	9.8236	998	996004	994011992	31.5911	9.9933
949	900601	854670349	30.8058	9.8270	999	998001	997002999	31.6070	9.9967
					1,000	1000000	1000000000	31.6229	10.0000

Table 36. Mensuration: The Measurement of Geometric Quantities.

MENSURATION

TRIANGLES

Area = base a times ½ altitude b.

QUADRILATERALS

Area = base a times altitude b.

Area = area of rectangle a plus area of triangles b, c, and d.

Area = triangle e + area of triangle f.

Area = area of rectangle b plus area of triangles b and c.

CIRCLE

Circumference of circle = 3.1416 times the diameter. 3 1/7 times the diameter, approximately: π times the diameter

(written πd): 6.2832 times the radius, or 2πr; 4 times the area divided by the diameter.

Area of circle = half the diameter multiplied by half the circumference: .7854 times the square of the diameter: 3.1416 times the square of half the diameter: square of the radius multiplied by 3.1416: circumference times the diameter divided by 4

Radius of circle = half the diameter: circumference multiplied by .159155: circumference divided by 6.2832: .564189 times the square root of the area.

Diameter of circle = twice the radius: circumference divided by 3.1416: circumference multiplied by .3183: 1.128 times the square root of the area.

Area of sector of circle = length of arc l times radius r; area of whole circle divided by 360 and multiplied by the number of degrees in the angle E. Square of the radius multiplied by the number of degrees in the angle E and by .00873.

Sector

Area of segment of circle = area of sector minus the area of the triangle formed by the chord and the two radii = ½ lr - [c(r-h)].

Segment

(Note: When either the angle E or the chord c is known the other may be found from a table of sines.)

Height h of segment of a circle =
$= r - \sqrt{r^2 - \frac{1}{4}c^2}$

Ring

Area of ring = area of the large circle D minus the area of the small circle d
$= \frac{\pi}{4}(D^2 - d^2).$

Ellipse

Area of ellipse = product of long diameter D and short diameter d multiplied by .7854.

Sphere

Area of surface of sphere = square of diameter multiplied by 3.1416.

Volume of sphere = cube of the diameter multiplied by .5236; or the cube of the radius multiplied by 4.1888.

Cylinder

Area of surface of cylinder = circumference multiplied by the height plus the area of both ends.

Volume of cylinder = area of base times the height or length; .7854 times square of diameter times height; 3.1416 times square of radius times height.

Circular Ring

Area of surface of circular link = 9.8696 times the mean diameter D times d.

Volume of circular link = 2.4674 times D times square of d.

Cone

Area of convex surface of cone = 1.5708 times diameter of base times slant height.

Total surface = convex surface plus .7854 times square of diameter.

Volume of cone = square of the diameter of the base multiplied by .7854 times one-third of the height.

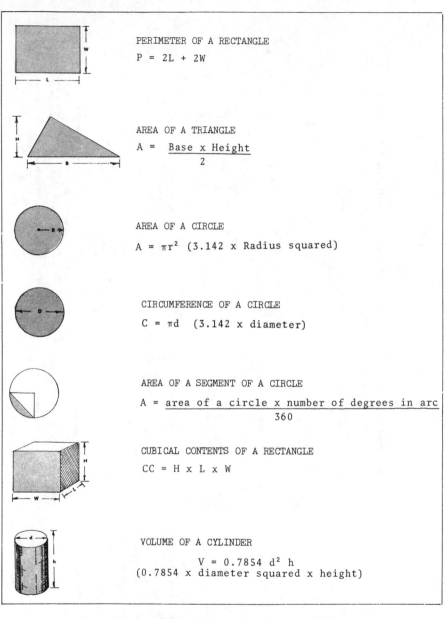

PERIMETER OF A RECTANGLE

P = 2L + 2W

AREA OF A TRIANGLE

$$A = \frac{\text{Base x Height}}{2}$$

AREA OF A CIRCLE

A = πr^2 (3.142 x Radius squared)

CIRCUMFERENCE OF A CIRCLE

C = πd (3.142 x diameter)

AREA OF A SEGMENT OF A CIRCLE

$$A = \frac{\text{area of a circle x number of degrees in arc}}{360}$$

CUBICAL CONTENTS OF A RECTANGLE

CC = H x L x W

VOLUME OF A CYLINDER

$$V = 0.7854 \ d^2 \ h$$
(0.7854 x diameter squared x height)

Figure 57. Simplified Solutions For Areas And Volumes.

Table 37. Matching Thermometer Scales.

Degrees Fahrenheit	Degrees Centigrade	Degrees Reaumer	Degrees Absolute	
			Fahrenheit	Centigrade
-460	-273.0	-218.0	0	0.0
-148	-100.0	-100.0	312	173.0
-50	-45.6	-36.4	410	227.4
-40	-40.0	-32.0	420	233.0
-30	-34.4	-27.5	430	238.6
-20	-28.9	-23.1	440	244.1
-10	-23.3	-18.6	450	249.7
0	-17.8	-14.2	460	255.2
10	-12.2	-9.8	470	260.8
20	-6.7	-5.3	480	266.3
30	-1.1	-0.9	490	271.9
32	0.0	0.0	492	273.0
40	4.5	3.5	500	277.5
50	10.0	8.0	510	283.0
60	15.6	12.4	520	288.6
70	21.1	16.9	530	294.1
80	26.7	21.3	540	299.7
90	32.2	25.8	550	305.2
100	37.8	30.2	560	310.8
110	43.3	34.0	570	316.3
120	48.9	39.1	580	321.9
130	54.4	43.3	590	327.4
140	60.0	48.0	600	333.0
150	65.6	52.4	610	338.6
160	71.1	56.9	620	344.1
170	76.7	61.3	630	349.7
180	82.2	65.7	640	355.2
190	87.8	70.2	650	360.8
200	93.3	74.5	660	366.3
210	98.9	79.0	670	371.9
212	100.0	80.0	672	373.0

Absolute Pressure	Gage Pressure	Atmospheres of Pressure	Inches Mercury Column	Centimeters Mercury Column	Feet of Water Column
0	30" vacuum	0	0	0	0
4.9	20" vacuum	0.33	9.95	25.4	11.3
9.8	10" vacuum	0.67	19.95	50.7	22.6
14.7*	Zero Pounds	1.0	29.92**	76.0	34.0
25.0	10.3 lbs.	1.7	50.7	129.3	57.7
30.0	15.3 lbs.	2.0	60.9	155.0	69.3
45.0	30.3 lbs.	3.1	91.3	233.0	104.0
60.0	45.0 lbs.	4.1	122.0	310.0	138.8
75.0	60.0 lbs.	5.1	152.0	388.0	173.0
100.0	85.0 lbs.	6.7	203.0	517.0	231.0
300.0	285.0 lbs.	20.0	609.0	1550.0	693.0
600.0	585.0 lbs.	40.1	1218.0	3100.0	1386.0
1200.0	1185.0 lbs.	81.7	2440.0	6200.0	2770.0
1500.0	1485.0 lbs.	102.0	3045.0	7750.0	3470.0

Table 38. Temperature Conversion Fahrenheit–Celsius (Centigrade).

459.4 to 25

C	value	F
−273	−459.4	
−268	−450	
−262	−440	
−257	−430	
−251	−420	
−246	−410	
−240	−400	
−234	−390	
−229	−380	
−223	−370	
−218	−360	
−212	−350	
−207	−340	
−201	−330	
−196	−320	
−190	−310	
−184	−300	
−179	−290	
−173	−280	
−169	−273	−459.4
−168	−270	−454
−162	−260	−436
−157	−250	−418
−151	−240	−400
−146	−230	−382
−140	−220	−364
−134	−210	−346
−129	−200	−328
−123	−190	−310
−118	−180	−292
−112	−170	−274
−107	−160	−256
−101	−150	−238
−96	−140	−220
−90	−130	−202
−84	−120	−184
−79	−110	−166
−73	−100	−148
−68	−90	−130
−62	−80	−112
−57	−70	−94
−51	−60	−76
−46	−50	−58
−40	−40	−40
−34	−30	−22
−29	−20	−4
−23	−10	14
−17.8	0	32
−17.2	1	33.8
−16.7	2	35.6
−16.1	3	37.4
−15.6	4	39.2
−15.0	5	41.0
−14.4	6	42.8
−13.9	7	44.6
−13.3	8	46.4
−12.8	9	48.2
−12.2	10	50.0
−11.7	11	51.8
−11.1	12	53.6
−10.6	13	55.4
−10.0	14	57.2
−9.4	15	59.0
−8.9	16	60.8
−8.3	17	62.6
−7.8	18	64.4
−7.2	19	66.2
−6.7	20	68.0
−6.1	21	69.8
−5.6	22	71.6
−5.0	23	73.4
−4.4	24	75.2
−3.9	25	77.0

26 to 99

C	value	F
−3.3	26	78.8
−2.8	27	80.6
−2.2	28	82.4
−1.7	29	84.2
−1.1	30	86.0
−0.6	31	87.8
0.0	32	89.6
0.6	33	91.4
1.1	34	93.2
1.7	35	95.0
2.2	36	96.8
2.8	37	98.6
3.3	38	100.4
3.9	39	102.2
4.4	40	104.0
5.0	41	105.8
5.6	42	107.6
6.1	43	109.4
6.7	44	111.2
7.2	45	113.0
7.8	46	114.8
8.3	47	116.6
8.9	48	118.4
9.4	49	120.2
10.0	50	122.0
10.6	51	123.8
11.1	52	125.6
11.7	53	127.4
12.2	54	129.2
12.8	55	131.0
13.3	56	132.8
13.9	57	134.6
14.4	58	136.4
15.0	59	138.2
15.6	60	140.0
16.1	61	141.8
16.7	62	143.6
17.2	63	145.4
17.8	64	147.2
18.3	65	149.0
18.9	66	150.8
19.4	67	152.6
20.0	68	154.4
20.6	69	156.2
21.1	70	158.0
21.7	71	159.8
22.2	72	161.6
22.8	73	163.4
23.3	74	165.2
23.9	75	167.0
24.4	76	168.8
25.0	77	170.6
25.6	78	172.4
26.1	79	174.2
26.7	80	176.0
27.2	81	177.8
27.8	82	179.6
28.3	83	181.4
28.9	84	183.2
29.4	85	185.0
30.0	86	186.8
30.6	87	188.6
31.1	88	190.4
31.7	89	192.2
32.2	90	194.0
32.8	91	195.8
33.3	92	197.6
33.9	93	199.4
34.4	94	201.2
35.0	95	203.0
35.6	96	204.8
36.1	97	206.6
36.7	98	208.4
37.2	99	210.2

100 to 820

C	value	F
38	100	212
43	110	230
49	120	248
54	130	266
60	140	284
66	150	302
71	160	320
77	170	338
82	180	356
88	190	374
93	200	392
99	210	410
100	212	413.6
104	220	428
110	230	446
116	240	464
121	250	482
127	260	500
132	270	518
138	280	536
143	290	554
149	300	572
154	310	590
160	320	608
166	330	626
171	340	644
177	350	662
182	360	680
188	370	698
193	380	716
199	390	734
204	400	752
210	410	770
216	420	788
221	430	806
227	440	824
232	450	842
238	460	860
243	470	878
249	480	896
254	490	914
260	500	932
266	510	950
271	520	968
277	530	986
282	540	1004
288	550	1022
293	560	1040
299	570	1058
304	580	1076
310	590	1094
316	600	1112
321	610	1130
327	620	1148
332	630	1166
338	640	1184
343	650	1202
349	660	1220
354	670	1238
360	680	1256
366	690	1274
371	700	1292
377	710	1310
382	720	1328
388	730	1346
393	740	1364
399	750	1382
404	760	1400
410	770	1418
416	780	1436
421	790	1454
427	800	1472
432	810	1490
438	820	1508

830 to 1540

C	value	F
443	830	1526
449	840	1544
454	850	1562
460	860	1580
466	870	1598
471	880	1616
477	890	1634
482	900	1652
488	910	1670
493	920	1688
499	930	1706
504	940	1724
510	950	1742
516	960	1760
521	970	1778
527	980	1796
532	990	1814
538	1000	1832
543	1010	1850
549	1020	1868
554	1030	1886
560	1040	1904
566	1050	1922
571	1060	1940
577	1070	1958
582	1080	1976
588	1090	1994
593	1100	2012
599	1110	2030
604	1120	2048
610	1130	2066
616	1140	2084
621	1150	2102
627	1160	2120
632	1170	2138
638	1180	2156
643	1190	2174
649	1200	2192
654	1210	2210
660	1220	2228
666	1230	2246
671	1240	2264
677	1250	2282
682	1260	2300
688	1270	2318
693	1280	2336
699	1290	2354
704	1300	2372
710	1310	2390
716	1320	2408
721	1330	2426
727	1340	2444
732	1350	2462
738	1360	2480
743	1370	2498
749	1380	2516
754	1390	2534
760	1400	2552
766	1410	2570
771	1420	2588
777	1430	2606
782	1440	2624
788	1450	2642
793	1460	2660
799	1470	2678
804	1480	2696
810	1490	2714
816	1500	2732
821	1510	2750
827	1520	2768
832	1530	2786
838	1540	2804

1550 to 2260

C	value	F
843	1550	2822
849	1560	2840
854	1570	2858
860	1580	2876
866	1590	2894
871	1600	2912
877	1610	2930
882	1620	2948
888	1630	2966
893	1640	2984
899	1650	3002
904	1660	3020
910	1670	3038
916	1680	3056
921	1690	3074
927	1700	3092
932	1710	3110
938	1720	3128
943	1730	3146
949	1740	3164
954	1750	3182
960	1760	3200
966	1770	3218
971	1780	3236
977	1790	3254
982	1800	3272
988	1810	3290
993	1820	3308
999	1830	3326
1004	1840	3344
1010	1850	3362
1016	1860	3380
1021	1870	3398
1027	1880	3416
1032	1890	3434
1038	1900	3452
1043	1910	3470
1049	1920	3488
1054	1930	3506
1060	1940	3524
1066	1950	3542
1071	1960	3560
1077	1970	3578
1082	1980	3596
1088	1990	3614
1093	2000	3632
1099	2010	3650
1104	2020	3668
1110	2030	3686
1116	2040	3704
1121	2050	3722
1127	2060	3740
1132	2070	3758
1138	2080	3776
1143	2090	3794
1149	2100	3812
1154	2110	3830
1160	2120	3848
1166	2130	3866
1171	2140	3884
1177	2150	3902
1182	2160	3920
1188	2170	3938
1193	2180	3956
1199	2190	3974
1204	2200	3992
1210	2210	4010
1216	2220	4028
1221	2230	4046
1227	2240	4064
1232	2250	4082
1238	2260	4100

2270 to 3000

C	value	F
1243	2270	4118
1249	2280	4136
1254	2290	4154
1260	2300	4172
1266	2310	4190
1271	2320	4208
1277	2330	4226
1282	2340	4244
1288	2350	4262
1293	2360	4280
1299	2370	4298
1304	2380	4316
1310	2390	4334
1316	2400	4352
1321	2410	4370
1327	2420	4388
1332	2430	4406
1338	2440	4424
1343	2450	4442
1349	2460	4460
1354	2470	4478
1360	2480	4496
1366	2490	4514
1371	2500	4532
1377	2510	4550
1382	2520	4568
1388	2530	4586
1393	2540	4604
1399	2550	4622
1404	2560	4640
1410	2570	4658
1416	2580	4676
1421	2590	4694
1427	2600	4712
1432	2610	4730
1438	2620	4748
1443	2630	4766
1449	2640	4784
1454	2650	4802
1460	2660	4820
1466	2670	4838
1471	2680	4856
1477	2690	4874
1482	2700	4892
1488	2710	4910
1493	2720	4928
1499	2730	4946
1504	2740	4964
1510	2750	4982
1516	2760	5000
1521	2770	5018
1527	2780	5036
1532	2790	5054
1538	2800	5072
1543	2810	5090
1549	2820	5108
1554	2830	5126
1560	2840	5144
1566	2850	5162
1571	2860	5180
1577	2870	5198
1582	2880	5216
1588	2890	5234
1593	2900	5252
1599	2910	5270
1604	2920	5288
1610	2930	5306
1616	2940	5324
1621	2950	5342
1627	2960	5360
1632	2970	5378
1638	2980	5396
1643	2990	5414
1649	3000	5432

In the center column, find the temperature to be converted. The equivalent temperature is in the left column if converting to Celsius, and in the right column if converting to Fahrenheit.

Table 39. Relative Humidity Scale.

DIFFERENCE BETWEEN THE DRY BULB AND WET BULB THERMOMETERS — DEGREES F

DRY BULB TEMPERATURE — DEGREES F

Dry bulb °F	1	2	3	4	5	6	7	8	9	10	11	12	13	14	15	16	17	18	19	20	21	22	23	24	25	26	27	28	29	30	31	32	33	34	35
31	89	78	68	58	47	37	28	18	8																										
32	89	79	69	59	49	39	30	20	11	2																									
33	90	80	70	60	51	41	32	23	14	5																									
34	90	81	71	62	52	43	34	25	16	8																									
35	91	81	72	63	54	45	36	27	19	10	2																								
36	91	82	73	64	55	46	38	29	21	13	5																								
37	91	83	74	65	57	48	40	31	23	15	7																								
38	91	83	75	66	58	50	42	33	25	17	10	2																							
39	92	83	75	67	59	51	43	35	27	20	12	5																							
40	92	83	75	68	60	52	45	37	29	22	15	7	0																						
41	92	84	76	69	61	54	46	39	31	24	17	10	3																						
42	92	85	77	69	62	55	47	40	33	26	19	12	5																						
43	93	85	77	70	63	56	48	42	35	28	21	14	8	1																					
44	93	85	78	71	63	56	49	43	36	30	23	16	10	4																					
45	93	86	78	71	64	57	51	44	38	31	25	18	12	6																					
46	93	86	79	72	65	58	52	45	39	32	26	20	14	8	2																				
47	93	86	79	72	66	59	53	46	40	34	28	22	16	10	5																				
48	93	86	79	73	66	60	54	47	41	35	29	23	18	12	7	1																			
49	93	86	80	73	67	61	54	48	42	36	31	25	19	14	9	3																			
50	93	87	80	74	67	61	55	49	43	38	32	27	21	16	10	5	0																		
51	94	87	81	75	68	62	56	50	45	39	34	28	23	17	12	7	2																		
52	94	87	81	75	69	63	57	51	46	40	35	29	24	19	14	9	4																		
53	94	87	81	75	69	63	58	52	47	41	36	31	26	22	17	12	8	3																	
54	94	88	82	76	70	64	59	53	48	42	37	32	27	22	18	13	9	5	0																
55	94	88	82	76	70	65	59	54	49	43	38	33	28	23	19	14	10	6	2																
56	94	88	82	76	71	65	60	55	50	44	39	34	30	25	20	16	11	7	4																
57	94	88	82	77	71	66	61	55	50	45	40	35	31	26	22	17	13	9	6	2															
58	94	88	83	77	72	66	61	56	51	46	41	37	32	27	23	18	14	11	7	4	1														
59	94	89	83	78	72	67	62	57	52	47	42	38	33	29	24	20	16	13	9	5	3														
60	94	89	83	78	73	68	63	58	53	48	43	39	34	30	26	21	17	14	10	7	4	1													
61	94	89	84	78	73	68	63	58	54	49	44	40	35	31	27	22	18	15	11	8	5	2													
62	94	89	84	79	74	69	64	59	54	50	45	41	36	32	28	24	20	16	13	9	7	4													
63	95	89	84	79	74	69	64	60	55	50	46	42	37	33	29	25	21	17	13	10	7	4	2												
64	95	90	84	79	74	70	65	60	56	51	47	43	38	34	30	26	22	18	15	11	7	6	3	0											
65	95	90	85	80	75	70	66	61	57	52	48	44	40	36	32	27	24	20	17	14	10	7	5	2											
66	95	90	85	80	75	71	66	61	57	53	49	45	41	37	32	29	25	21	18	14	11	9	6	3	0										
67	95	90	85	80	76	71	67	62	58	53	50	46	42	38	34	30	26	23	19	16	13	10	6	4	2										
68	95	90	85	80	76	71	67	62	58	54	50	46	42	38	35	31	27	24	21	18	14	11	8	6	3	1									
69	95	90	85	81	76	72	67	63	59	55	51	47	43	39	36	32	28	25	22	19	15	12	10	7	5	3									
70	95	90	86	81	77	72	68	64	60	55	52	48	44	40	37	33	30	26	23	20	17	14	11	9	6	4	1								
71	95	91	86	82	77	73	69	64	60	56	52	48	45	41	38	34	30	27	24	21	18	15	13	10	7	5	3								
72	95	91	86	82	78	73	69	65	61	57	53	49	46	42	39	35	32	28	25	22	19	17	13	11	9	6	4	1							
73	95	91	86	82	78	74	70	65	61	57	53	50	46	43	39	36	33	29	26	23	21	18	15	13	10	8	5	3							
74	95	91	86	82	78	74	70	66	62	58	54	50	47	43	40	37	33	30	27	24	20	17	15	11	8										
75	96	91	86	82	78	74	70	66	62	58	54	51	47	44	41	37	34	31	28	25	22	18	16	13	11	9	6	4	1						
76	96	91	87	82	78	74	70	66	62	59	55	51	48	44	41	38	35	31	28	26	23	19	16	14	12	9	7	5	3						
77	96	91	87	83	79	75	71	67	63	60	56	52	48	45	42	39	36	32	29	27	24	21	18	16	13	10	8	6	4						
78	96	91	87	83	79	75	71	67	64	60	56	53	49	46	43	40	37	33	30	28	25	22	20	17	14	12	10	8	5						
79	96	92	87	83	80	76	72	68	65	60	57	53	50	47	44	40	37	35	32	29	26	23	21	18	15	13	11	9	6	4					
80	96	92	88	84	80	76	72	68	65	61	57	54	50	47	44	41	38	35	32	30	27	24	22	20	17	14	12	10	7	5	2				
82	96	92	88	84	80	77	73	69	66	62	59	55	52	48	46	43	40	37	34	31	29	26	24	21	19	17	15	13	11	8	6	3			
84	96	92	88	84	81	77	74	70	67	63	60	57	53	50	47	44	41	38	35	33	30	28	25	23	20	18	16	14	11	9	7	5	3		
86	96	92	88	84	81	77	74	70	67	64	61	57	54	51	48	45	42	40	37	33	31	29	26	24	22	20	18	16	13	11	9	7	5	3	1
88	96	92	88	85	81	78	74	71	68	64	61	58	55	52	49	46	43	40	37	35	32	30	27	25	23	21	19	17	15	12	10	7	5	3	1
90	96	92	89	85	81	78	75	71	68	65	62	59	56	53	50	48	45	42	40	37	35	32	30	28	25	23	21	19	17	15	13				
92	96	92	89	85	82	79	75	72	69	66	63	60	57	54	51	48	45	43	40	38	35	33	30	28	26	24	22	20	18	16	14				
94	96	93	89	85	82	79	76	72	69	66	63	60	57	54	51	49	46	43	41	38	36	34	31	29	27	25	23	21	19	17	15				
96	96	93	89	86	82	79	76	73	70	66	63	61	58	55	52	50	47	44	42	39	37	35	32	30	28	26	24	22	20	18	16				
98	96	93	89	86	83	80	77	73	70	67	64	61	58	56	53	50	48	45	43	40	38	36	33	31	29	27	25	23	21	19	17				
100	96	93	90	86	83	80	77	74	70	67	64	62	59	56	54	51	48	46	44	41	39	37	34	32	30	28	26	24	22	20	18				
102	96	93	90	86	83	80	77	74	71	68	65	62	59	57	54	52	49	47	44	42	40	38	35	33	31	29	27	25	23	21	19				
104	97	93	90	87	83	81	77	74	71	68	65	63	60	57	55	52	50	47	45	42	40	38	36	34	32	30	28	26	24	22	20				
106	97	93	90	87	84	81	78	75	72	69	66	63	60	58	55	53	50	48	46	43	41	39	37	35	33	31	29	27	25	23	21				
108	97	93	90	87	84	81	78	75	72	69	66	64	61	58	56	54	51	49	46	44	42	40	38	36	34	32	30	28	26	24	22				
110	97	93	90	87	84	81	78	75	72	69	67	64	61	59	56	54	52	49	47	45	43	41	39	37	35	33	31	29	27	25	23				
112	97	94	90	87	84	82	79	76	73	70	67	65	62	60	57	55	53	51	47	46	44	42	40	38	36	34	32	30	28	26	24				
114	97	94	91	88	85	82	79	76	74	71	68	66	63	61	58	56	54	52	49	47	45	43	41	39	37	35	33	31	29	27	25	24	22		
116	97	94	91	88	85	83	80	77	74	72	69	66	64	61	59	57	55	52	50	48	46	44	42	40	38	36	34	33	31	29	27	25	23		
118	97	94	91	88	86	83	80	77	75	72	69	67	64	62	60	58	55	53	51	49	47	45	43	41	39	37	35	34	32	30	28	27	25		
120	97	94	91	88	86	83	80	78	75	72	70	67	65	63	60	58	56	54	52	50	48	46	44	42	40	38	37	35	33	31	29	28	26		
122	97	94	91	88	86	83	81	78	75	73	70	68	65	63	61	59	57	55	53	51	49	47	45	43	41	39	38	36	34	32	31	29	27		
124	97	94	91	89	86	84	81	78	76	73	71	68	66	64	62	60	57	55	53	51	50	48	46	44	42	40	39	37	35	33	32	30	28		
126	97	94	91	89	86	84	81	79	76	74	71	69	66	64	62	60	58	56	54	52	50	49	47	45	43	41	40	38	36	35	33	31	29		
128	97	94	92	89	87	84	82	79	77	74	72	69	67	65	63	61	59	57	55	53	51	50	48	46	44	43	41	39	38	36	34	32	30		
130	97	94	92	89	87	85	82	80	77	75	72	70	68	65	63	61	59	58	56	54	52	50	49	47	45	43	42	40	39	37	35	33	31		
132	97	94	92	89	87	85	82	80	78	75	73	70	68	66	64	62	60	58	56	54	53	51	49	48	46	44	43	41	39	38	36	34	32		

RELATIVE HUMIDITY

Example: Dry bulb temperature is 90 F, if the wet bulb temperature is 82 F, read below the 8 degree wet bulb depression column to find the percent relative humidity opposite 90 F = 71% rh.

Table 40. Free Area Openings For Common Restrictive Devices.

MATERIAL	PERCENT FREE AREA
Wood Louvers – doors & grilles	20 – 25
Insect Screens	40 – 50
Metal Grilles – supply & return	60 – 75
Dust Filters	60 – 80
Wire Cloth – bird screens	70 – 90

Note: Restricting air devices are often found on existing jobs—and sometimes on new construction—where the balancing technician does not have access to manufacturer's data. Such data may have been undetermined, unpublished, not entered in the job file, or simply lost. In such cases an approximation can often be of great help. Using the above table for example, for a 6'-0" wood full louvered door, the net free area would be somewhere between 3.6 and 4.5 sq. ft. The factors shown are based on clean surfaces and vary with the design of the device.

Table 41. Grains Of Moisture Per Cubic Foot Of Air At Various Temperatures.

% RELATIVE HUMIDITY	TEMPERATURE – DEGREES F										
	75	72	70	67	65	60	50	40	30	20	10
100	9.35	8.51	7.98	7.24	6.78	5.74	4.08	2.85	1.94	1.23	0.78
90	8.42	7.66	7.18	6.52	6.10	5.17	3.67	2.56	1.74	1.11	0.70
80	7.49	6.81	6.38	5.79	5.43	4.60	3.26	2.28	1.55	0.99	0.62
70	6.55	5.96	5.59	5.07	4.75	4.02	2.85	1.99	1.35	0.86	0.54
60	5.61	5.11	4.79	4.35	4.07	3.45	2.45	1.71	1.16	0.74	0.47
50	4.68	4.25	3.99	3.62	3.39	2.87	2.04	1.42	0.97	0.62	0.39
40	3.74	3.40	3.19	2.90	2.71	2.30	1.63	1.14	0.78	0.49	0.31
30	2.81	2.55	2.39	2.17	2.04	1.72	1.22	0.86	0.58	0.37	0.23
20	1.87	1.70	1.60	1.45	1.36	1.15	0.82	0.57	0.39	0.25	0.16
10	0.94	0.85	0.80	0.72	0.68	0.57	0.41	0.29	0.19	0.12	0.08

Table 42. Wind Forces At Common Velocities.

MILES PER HOUR MPH	METERS PER SECOND M/SEC	FEET PER MINUTE FPM	POUNDS PER SQUARE FEET PSF	INCHES OF WATER IN. H_2O
2	0.9	176	0.010	0.002
3	1.3	264	0.025	0.004
4	1.8	352	0.040	0.007
5	2.2	440	0.062	0.012
6	2.7	528	0.090	0.017
7	3.1	616	0.12	0.023
8	3.7	704	0.16	0.030
9	4.0	792	0.20	0.039
10	4.5	880	0.25	0.048
12	5.4	1056	0.36	0.069
14	6.3	1232	0.49	0.094
16	7.2	1408	0.64	0.123
18	8.1	1584	0.81	0.155
20	9.0	1760	1.00	0.192
22	9.8	1936	1.21	0.232
24	10.7	2112	1.44	0.276
26	11.6	2288	1.69	0.324
28	12.5	2464	1.96	0.376
30	13.4	2640	2.25	0.432
32	14.3	2816	2.56	0.491
34	15.2	2992	2.89	0.555
36	16.1	3168	3.24	0.622
38	17.0	3344	3.61	0.693
40	17.9	3520	4.00	0.768

Multiply	By	To Obtain
miles per hour	88	feet per minute
miles per hour	0.447	metres per second
feet per minute	0.01136	miles per hour
metres per second	2.237	miles per hour

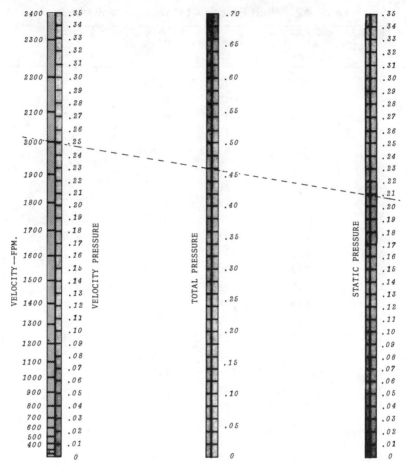

Figure 58. Speed-O-Graph: CONVERSION: Velocity, Static Pressure, Velocity Pressure, Total Pressure.

Given: Duct velocity 2000 fpm; static pressure, 0.21 in water.

Find: Velocity pressure and total pressure.

1. Lay a straightedge across the two outer scales intersecting 0.21 on the Sp scale and 2000 on the velocity scale. 2. Read the answer = 0.25 Vp and 0.46 Tp. *Note:* When laying the straightedge, cross the scales on the heavy center vertical line.

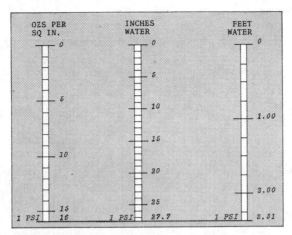

Figure 58A. *Pressure Equivalents.* Fan static pressures are usually given in terms of inches of water rather than psi because a small decimal such as 0.036 psi is more combersome to work with than its equivalent, 1.0 in. of water. Pressure and head are equivalent terms; 1 psi corresponds to 27.7 in. of water.

MULTIPLY	X	TO OBTAIN
Psi	16	$Oz/in.^2$
Psi	2.31	Ft H_2O
Psi	27.73	In. H_2O
Psi	0.0703	Kg/cm^2
Psi	2.036	In. Hg
In. H_2O	0.07342	In. Hg
In. H_2O	0.5770	$Oz/In.^2$
In. H_2O	0.03606	Psi
In. H_2O	5.196	Psf
Ft H_2O	0.4328	Psi
Ft H_2O	62.32	Psf
TO OBTAIN ABOVE	DIVIDE ABOVE	STARTING WITH ABOVE

$Oz/in.^2$ = ounces per square inch
in. Hg. = inches of mercury @ 0 C
ft H_2O = feet of water @ 68 F

in. H_2O = inches of water @ 68 F
Psi = pounds per square inch
Psf = pounds per sqare foot

14.7 psi
29.92 in. Hg }= one atmosphere
33.9 ft water

The formula for converting vacuum gage readings to absolute pressure is,

$$\frac{30 \text{ minus psig vacuum}}{2} = \text{psia} \qquad (50)$$

Absolute pressure equals gage pressure plus 14.7

See also, pressure conversion factors in Table 31.

Figure 58B. Pressure Conversion Factors

187

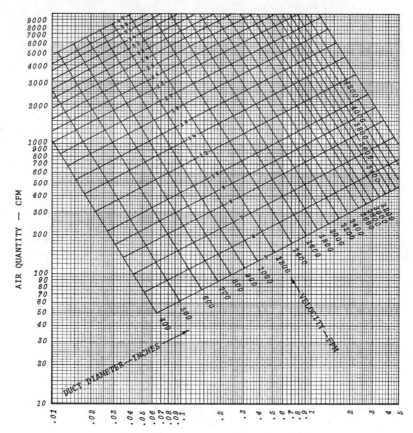

FRICTION LOSS — INCHES OF WATER PER 100 FT EQUIVALENT LENGTH

Figure 59. Speed-O-Graph: Friction Loss For Round Duct.

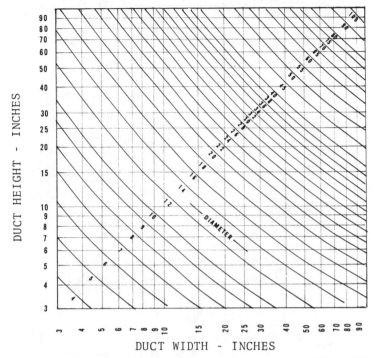

DUCT HEIGHT - INCHES

DUCT WIDTH - INCHES

Figure 60. Speed-O-Graph: Rectangular Equivalents Of Round Ducts.

The equivalent duct dimension for a rectangular duct with sides of dimensions a and b is

$$\sqrt{\frac{4ab}{\pi}}$$

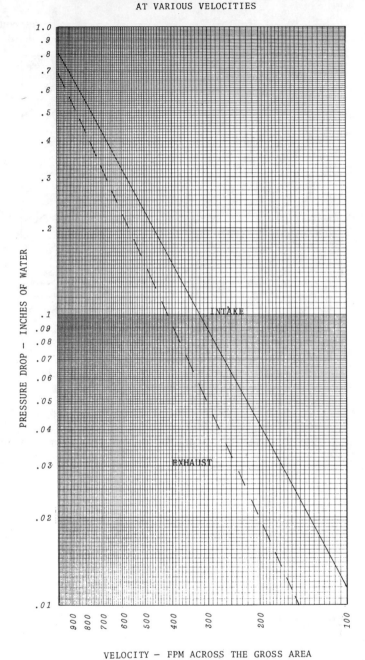

VELOCITY — FPM ACROSS THE GROSS AREA

Figure 61. Speed-O-Graph: Pressure Drop Of Intake And Exhaust Grilles At Various Velocities.

On existing jobs, and sometimes on new construction, manufacturer's data for grilles, etc. may not be available. Where such data are not available this chart may be used to determine approximate conditions. The chart gives *average* pressure drop curves and is not intended as a substitute for specific manufacturer's data.

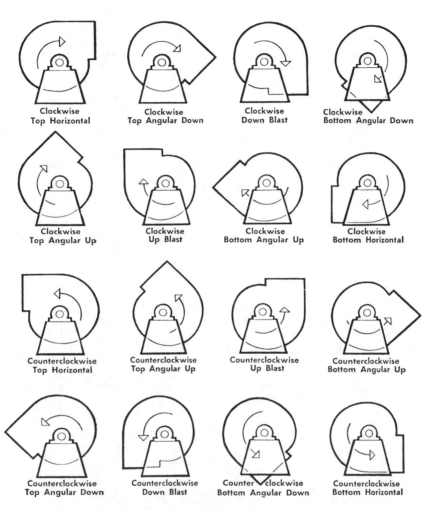

Direction of rotation is determined from drive side of fan.
On single inlet fans, drive side is always considered as the side opposite fan inlet.
On double inlet fans with drives on both sides, drive side is that with the higher powered drive unit.
Direction of discharge is determined in accordance with diagrams. Angle of discharge is referred
    to the horizontal axis of fan and designated in degrees above or below such standard reference axis.
For fan inverted for ceiling suspension, or side wall mounting, direction of rotation and discharge is
    determined when fan is resting on floor.

| Adopted 5-26-60 | DESIGNATIONS FOR ROTATION AND DISCHARGE | AMCA STANDARD |
| Revised 1-30-68 | OF CENTRIFUGAL FANS | 2406-66 |

Figure 62. AMCA Fan Standard.

SW — Single Width   DW — Double Width
SI — Single Inlet   DI — Double Inlet

Arrangements 1, 3, 7 and 8 are also available with bearings mounted on pedestals or base set independent of the fan housing.

For designation of rotation and discharge,   see AS 2406.
For motor position, belt or chain drive,   see AS 2407.
For designation of position of inlet boxes,   see AS 2405.

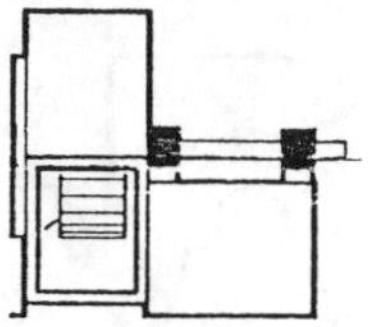

ARR. 1 SWSI For belt drive or direct connection. Impeller overhung. Two bearings on base.

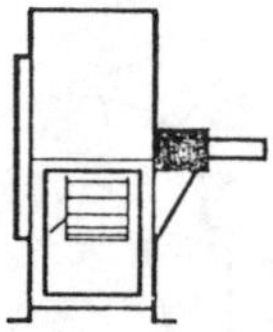

ARR. 2 SWSI For belt drive or direct connection. Impeller overhung. Bearings in bracket supported by fan housing.

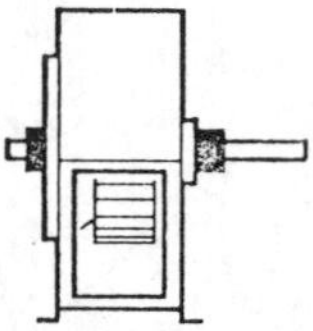

ARR. 3 SWSI For belt drive or direct connection. One bearing on each side and supported by fan housing.

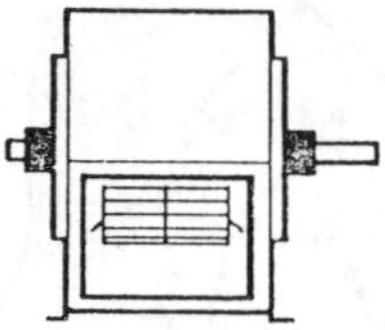

ARR. 3 DWDI For belt drive or direct connection. One bearing on each side and supported by fan housing.

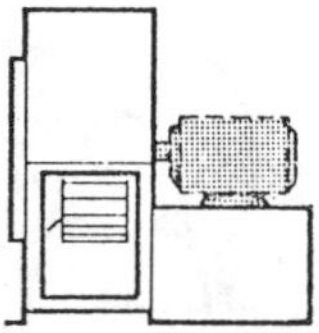

ARR. 4 SWSI For direct drive. Impeller overhung on prime mover shaft. No bearings on fan. Prime mover base mounted or integrally directly connected.

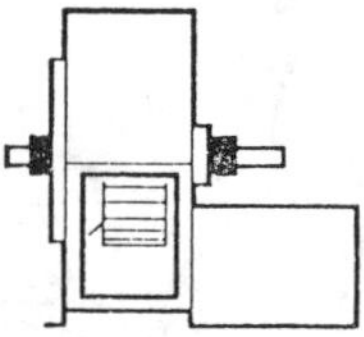

ARR. 7 SWSI For belt drive or direct connection. Arrangement 3 plus base for prime mover.

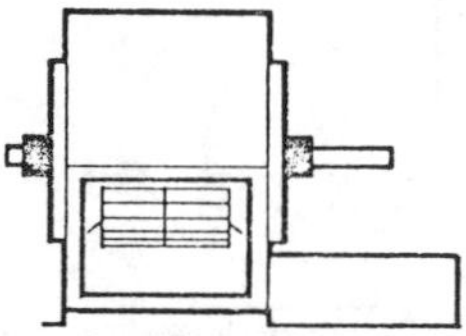

ARR. 7 DWDI For belt drive or direct connection. Arrangement 3 plus base for prime mover.

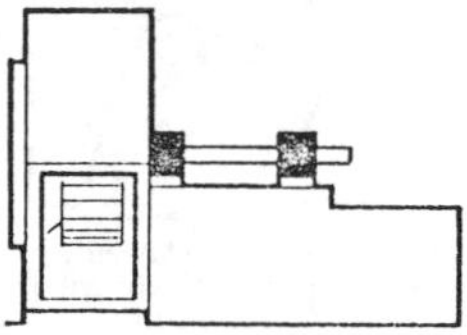

ARR. 8 SWSI For belt drive or direct connection. Arrangement 1 plus extended base for prime mover.

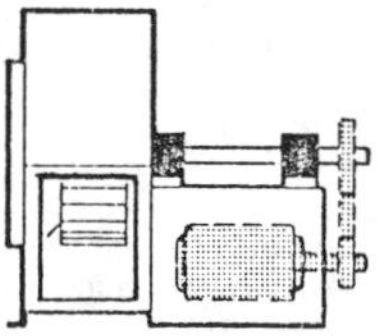

ARR. 9 SWSI For belt drive. Impeller overhung, two bearings, with prime mover outside base.

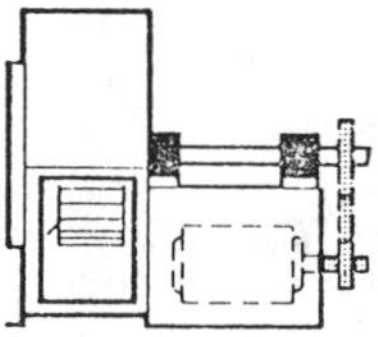

ARR. 10 SWSI For belt drive. Impeller overhung, two bearings, with prime mover inside base.

DRIVE ARRANGEMENTS FOR CENTRIFUGAL FANS

AMCA STANDARD 2404-72

Adopted 4-1-45 Revised 10-30-72

Figure 63. AMCA Fan Standard.

Table 43. Saturated Refrigerant Temperature–Pressure Chart.

Italicized figures are inches of mercury;
bold type is gauge pressure in lbs per sq in.

Temp. °F	R-717 (Ammonia)	R-11	R-12	R-22	R-500	R-502
	psig	psig	psig	psig	psig	psig
-70	*21.9*	*29.5*	*21.8*	*16.6*	*20.3*	*12.6*
-65	*20.4*	*29.3*	*20.5*	*14.4*	*18.8*	*10.0*
-60	*18.6*	*29.2*	*19.0*	*12.0*	*17.0*	*7.0*
-55	*16.6*	*29.0*	*17.3*	*9.2*	*15.0*	*3.6*
-50	*14.3*	*28.9*	*15.4*	*6.2*	*12.8*	0.0
-45	*11.7*	*28.7*	*13.3*	*2.7*	*10.4*	2.1
-40	*8.7*	*28.4*	*10.9*	0.5	*7.6*	4.3
-35	*5.4*	*28.1*	*8.3*	2.6	*4.6*	6.7
-30	*1.6*	*27.8*	*5.5*	4.8	*1.2*	9.4
-25	1.3	*27.4*	*2.3*	7.4	1.2	12.3
-20	3.6	*27.0*	0.6	10.2	3.2	15.5
-15	6.2	*26.5*	2.5	13.2	5.4	19.0
-10	9.0	*26.0*	4.5	16.5	7.8	22.8
-5	12.2	*25.4*	6.7	20.1	10.4	26.7
0	15.7	*24.7*	9.2	24.0	13.3	31.2
5	19.6	*23.9*	11.8	28.2	16.4	36.0
10	23.8	*23.1*	14.6	32.8	19.7	41.1
15	28.4	*22.1*	17.7	37.7	23.4	46.6
20	33.5	*21.1*	21.0	43.0	27.3	52.5
25	39.0	*19.9*	24.6	48.8	31.5	58.7
30	45.0	*18.6*	28.5	54.9	36.0	65.4
35	51.6	*17.2*	32.6	61.5	40.9	72.6
40	58.6	*15.6*	37.0	68.5	46.1	80.2
45	66.3	*13.9*	41.6	76.0	51.7	88.3
50	74.5	*12.0*	46.7	84.0	57.6	96.9
55	83.4	*10.0*	52.0	92.6	63.9	106.0
60	92.9	*7.8*	57.7	101.6	70.6	115.6
65	103.1	*5.4*	63.8	111.2	77.8	125.8
70	114.1	*2.8*	70.2	121.4	85.4	136.6
75	125.8	0.0	77.0	132.3	93.5	148.0
80	138.3	1.5	84.2	143.6	102	159.9
85	151.7	3.2	91.6	155.7	111	172.6
90	165.9	4.9	99.8	168.4	121	185.8
95	181.1	6.8	108.3	181.8	131	199.8
100	197.2	8.8	117.2	195.9	141	214.4
105	214.2	10.9	126.6	210.8	153	229.8
110	232.3	13.2	136.4	226.4	164	245.8
115	251.5	15.6	146.8	242.7	176	262.7
120	271.7	18.3	157.6	259.9	189	280.3

Table 44. Formulas For Solving Refrigeration Problems.

Net Refrigerating Effect, Btu/lb	= Enthalpy of Vapor Leaving Evaporator, Btu/lb	− Enthalpy of Liquid Entering Evaporator, Btu/lb

$$\text{Compression Work, Btu/min} = \text{Heat of Compression, Btu/lb} \times \text{Refrigerant Circulated, lb/min}$$

$$\text{Compression Horsepower} = \frac{\text{Compression Work, Btu/min}}{42.4}$$

$$\text{Compression Horsepower} = \frac{\text{Capacity, Btu/min}}{42.4 \times \text{COP}}$$

$$\text{Compression Horsepower per Ton} = \frac{4.715}{\text{Coefficient of Performance}}$$

$$\text{Power, watts} = \text{Compression Horsepower per Ton} \times 745.7$$

$$\text{Coefficient of Performance} = \frac{\text{Net Refrigerating Effect, Btu/lb}}{\text{Heat of Compression, Btu/lb}}$$

$$\text{Capacity, Btu/min} = \text{Refrigerant Circulated, lb/min} \times \text{Net Refrigerating Effect, Btu/lb}$$

$$\text{Compressor Displacement, ft}^3/\text{min} = \frac{\text{Capacity, Btu/min} \times \text{Volume of Gas Entering Compressor, ft}^3/\text{lb}}{\text{Net Refrigerating Effect, Btu/lb}}$$

Heat of Compression, Btu/lb	= Enthalpy of Vapor Leaving Compressor, Btu/lb	− Enthalpy of Vapor Entering Compressor, Btu/lb

$$\text{Volumetric Efficiency} = 100 \times \frac{\text{Actual Weight of Refrigerant}}{\text{Theoretical Weight of Refrigerant}}$$

$$\text{Compression Ratio} = \frac{\text{Head Pressure, psia (absolute)}}{\text{Suction Pressure, psia (absolute)}}$$

$$\text{Refrigerant Circulated, lb (min) (ton)} = \frac{200}{\text{Refrigerating Effect}}$$

42.4 = heat flow, Btu/(min) (hp); 200 = Btu/(min) (ton); COP = coefficient of performance

Index